T0256980

NATURAL RESOURCES: *Quality and Quantity*

NATURAL RESOURCES: *Quality and Quantity:* PAPERS PRESENTED BEFORE A FACULTY SEMINAR AT THE UNIVERSITY OF CALIFORNIA, BERKELEY, 1961–1965

Edited by S. V. CIRIACY-WANTRUP *and* JAMES J. PARSONS *and published by the* UNIVERSITY OF CALIFORNIA PRESS, BERKELEY AND LOS ANGELES, 1967

University of California Press
Berkeley and Los Angeles, California
Cambridge University Press
London, England
Copyright © 1967, by
The Regents of the University of California
Library of Congress Catalog Card Number: 67-19220

CONTENTS

CONTRIBUTORS

WILLIAM A. ALBRECHT is emeritus professor of soils at the University of Missouri.

HAROLD H. BISWELL is professor of forestry at the University of California, Berkeley.

MICHAEL F. BREWER is vice-president, Resources for the Future, Inc., Washington, D. C. He was assistant professor of agricultural economics, University of California, Berkeley, when his paper was presented.

S. V. CIRIACY-WANTRUP is professor of agricultural economics at the University of California, Berkeley, and chairman of the Chancellor's Committee on Natural Resources.

EVERETT D. HOWE is professor of mechanical engineering and director of the Sea Water Conservation Project at the University of California, Berkeley.

A. STARKER LEOPOLD is professor of zoology at the University of California, Berkeley.

ALBERT LEPAWSKY is professor of political science at the University of California, Berkeley.

DANIEL B. LUTEN is lecturer in geography at the University of California, Berkeley.

J. BRIAN MUDD is associate professor of biochemistry and research associate in the Air Pollution Research Center, University of California, Riverside.

LEWIS MUMFORD, author and critic, lives in Amenia, New York. He was Visiting Research Professor of Governmental Affairs at the University of California, Berkeley, when his paper was presented.

JAMES J. PARSONS is professor of geography at the University of California, Berkeley, and vice-chairman of the Committee on Natural Resources.

WILLIAM PETERSEN is with the Institute of Human Sciences, Boston College. He was professor of sociology at the University of California, Berkeley, when his paper was presented.

MILNER B. SCHAEFER is professor of marine resources and director of the Institute of Marine Resources, University of California, San Diego.

ARNOLD M. SCHULTZ is professor of forestry at the University of California, Berkeley.

CURT STERN is professor of zoology at the University of California, Berkeley.

BERNARD D. TEBBENS is professor of industrial hygiene engineering in the School of Public Health at the University of California, Berkeley.

INTRODUCTION: BACKGROUND AND
OBJECTIVES   *S. V. Ciriacy-Wantrup*

The essays composing this volume were first presented informally before an interdisciplinary faculty seminar at the University of California, Berkeley. The seminar was sponsored by the Chancellor's Committee on Natural Resources.[1] A grant from the Conservation Foundation was helpful in conducting the seminar, especially in obtaining
speakers from outside the University.

The seminar met every two months on the average from 1961 to
1965. During this period many more papers were presented and
many more subjects dealt with than appear here. Some of the most
stimulating presentations had to be left out because of the difficulty of
persuading a busy faculty and guest speakers from other parts of the
country and the world to transform an informal evening's discussion
into formal papers. This explains some gaps in subjects covered, ap-
proaches used, and disciplines represented.

The study of the relations among man, society, and natural resources concerns many disciplines, whereas modern science fosters
and even requires specialization. As a consequence, the old barriers
between traditional disciplines have increased and new ones have
been erected between subdisciplines. The academic structure of the
modern American university emphasizes the department or school organized on the basis of a discipline or a subdiscipline. At the University of California at Berkeley, this academic structure is especially
well developed.

The effects of specialization and of an academic structure based on

---

[1] The members of the Committee were H. G. Baker (Botany), H. H. Biswell
(Forestry), A. H. Dupree (History), W. Eberhard (Sociology), E. D. Howe
(Engineering), A. S. Leopold (Zoology), A. Lepawsky (Political Science), J.
J. Parsons (Geography), vice-chairman, Sho Sato (Law), and S. V. Ciriacy-
Wantrup (Agricultural Economics), chairman.

disciplines upon the advancement of interdisciplinary areas such as the relations between man, society, and natural resources were the main concerns of the Committee. The Committee concluded that a faculty seminar with a membership carefully selected on an inter-disciplinary basis might stimulate and facilitate communication among the faculty of different disciplines. It was hoped that such a seminar might serve as a catalyst for bringing together faculty members from different departments for the planning of interdisciplinary courses. Finally, the seminar was to explore the possibilities for organizing interdisciplinary research on natural resources into an In-stitute of Natural Resources.

A continuing central theme was needed—one vital for the study of natural resources, broad enough to include several disciplines, rela-tively unexplored, and complex enough to challenge and hold the in-terests of mature scholars. The problems of quality and of the rela-tions between quality and quantity seemed especially promising. These problems are significant in many disciplines. In the study of natural resources, they had been neglected in the past. The difficulties in dealing with problems of quality, even in identifying them concep-tually, are great.

In retrospect, the choice of the main theme was a good one from the standpoint of the immediate objectives of the seminar. Faculty in-terest was widespread among disciplines, intense at nearly every meeting, and sustained over five years. From the standpoint of formal publication, however, reactions to the choice of the main theme and its treatment will be mixed. The specialist who looks for new devel-opments and definite conclusions in his own field will not be fully sat-isfied. The general reader will encounter some snags because of the technical terminology used and the special knowledge required, but he will benefit from the fact that highly qualified specialists were addressing an academic audience among whom their own specialty constituted a minority. Specialists who have explored their own field deeply enough to become aware of the necessity of relating it to other fields concerned with natural resources will be challenged to explore further on unfamiliar ground.

In conducting the seminar, each speaker was free to develop the main theme of "quality and quantity" as he saw it in his own area of competence. The resulting papers inevitably show considerable di-vergence, for the interpretation of the theme and of the concept "quality" itself differs among the authors. The common denominator, the overall objective of the seminar, was to explore the multidimen-sionality of natural resources and their use by man, rather than to deepen a conceptual dichotomy between quality and quantity. The main theme was quality *and* quantity, rather than quality *versus* quantity.

Some qualitative dimensions of resources and of resource use are less obvious and more difficult to measure with precision than others. This is one major reason why these dimensions have been relatively neglected in a scientific environment oriented toward precise quantitative measurement and why the emphasis in the seminar was placed on these dimensions. But some of the semantic aspects of the words "quality" and "quantity," like those of many dichotomies, are unfortunate, and some possible confusion must be guarded against.

Some dimensions of quality are no less amenable to precise quantitative measurement than those of quantity. For example, the quality dimensions of water—temperature, dissolved oxygen, biochemical oxygen demand, total dissolved solids, and, for individual water uses, such particular solids as chlorides, toxins, and synthetic organics— can be measured as precisely as volume, weight, or flow rate. Similar measures for dimensions of quality are commonly employed for other "composites" of resources—for instance, soil and air.[2]

Serious difficulties of quantitative measurement arise when these and similar physical dimensions of quality are evaluated by the social sciences and the humanities. Here also one must consider various dimensions of evaluation. If evaluation is one-dimensional, that is, for example, in pecuniary terms, for the purpose of optimizing social welfare in economics, problems both of validity and of relevance are present because other dimensions are left out. These other dimensions are usually significant for decisions of public policy. Under these conditions, a mere increase in the precision of quantitative measurement of only one dimension of evaluation becomes meaningless or misleading. The problem of the relevant degree of quantification in science is encountered in all attempts at defining social optima, and has been discussed elsewhere.[3] The first step in the approach to this problem is to identify the relevant dimensions of quality and of evaluation.

---

[2] For a more detailed discussion of a classification of resources, and of measuring their various dimensions, see S. V. Ciriacy-Wantrup, *Resource Conservation: Economics and Policies* (2d ed. rev.; Berkeley: Division of Agricultural Sciences, University of California, 1963), chap. 3.

[3] S. V. Ciriacy-Wantrup, "Water Policy," in Ven Te Chow, ed., *Handbook of Applied Hydrology: A Compendium of Water-Resource Technology* (New York: McGraw-Hill Book Co., 1964), sec. 28, pp. 28-1 to 28-25, and "Water Quality, A Problem for the Economist," *Journal of Farm Economics,* Vol. 63, No. 5 (December, 1961), pp. 1133–1144 [University of California, Giannini Foundation Paper No. 212].

The three essays of Part I deal with this issue of identification. Mumford is concerned with the quality of civilizations, Luten with the quality of the landscape, and Stern with the qualitative aspects of the population problem. In these three topics, the difficulties of identifying the dimensions of quality are of a different order of magnitude from those involved in identifying the dimensions of water quality. The difference is that evaluation is involved.

The seven essays of Part II are more concerned with the kind of quality dimensions illustrated by water quality. For these quality dimensions, the state of the arts has mastered quantitative measurement; and resource management is making a beginning in the consideration of quality issues. Still, the problems of evaluation and of identifying other than the physical dimensions of quality remain. It is implicitly touched upon in most essays of Part II, especially in the last one by Leopold, and becomes explicit in Part III.

To do justice to the four essays in Part III, it is only fair to their authors to state that the editors were responsible for determining the grouping of essays in the volume. Their decision was made *ex post,* after all papers were available, and the authors of Part III were not aware that their essays would have to carry the burden of pointing out directions for research and policy.

All essays of Part III were written by members of disciplines—one ecologist and three social scientists—dealing with complex biological-social systems dominated by the two great evolutionary forces of mutation and selection. It may not be entirely an accident nor the disciplinary bias of the editors that these are the disciplines in which the directions for research and policy are being developed. One may submit that the direction for research and policy in the area with which this volume is concerned is in assessing the functional significance of quality dimensions in the structure and performance of such biological–social systems. It is in this assessment that the problems of identification and of evaluation of quality merge.

Part 1   SOME BASIC CONSIDERATIONS

QUALITY IN THE CONTROL OF
QUANTITY *Lewis Mumford*

The problem of our age, which dominates or underlies most of its other problems, is the problem of quantity. The evidence increasingly shows that many technological innovations that have been regarded as indications of progress have actually been regressive in their effects, in that they have favored mechanical processes at the expense of human functions and human purposes. My purpose is to expose the ideological misconceptions that have impelled us to promote the quantative expansion of knowledge, power, productivity, without inventing any adequate system of controls. I propose further to show that the deliberate disregard of qualities in the conceptual scheme of seventeenth-century science has removed the complex, built-in controls which, though sometimes clumsy, had hitherto been effective.

The arts and crafts in every form, including engineering, had a long life, coterminous with civilization itself, before science began to influence them, and embedded in this deep compost of technological experience was a large body of accurate empirical observations: protoscience, we might call it. Knowledge of physical, chemical, and biological transformations, applied in metallurgy, pottery, brewing, dyeing, plant selection, animal breeding, and medicine was abundant and, as far as it went, sound, though it contained a turbid residue of confused observation, misinterpreted information, and effete magic. Two attributes distinguished it from the corpus of scientific knowledge that has largely supplanted it: this knowledge was coupled to practice, and it was attached not merely to causal insight into the means, but also to teleological understanding of ends; so that at no point was it wholly detached from its social matrix or its human destination.

This prescientific technology is still often regarded with contempt,

7

both on practical and on theoretic grounds. As to the first, ancient technics seem faulty because many material improvements that now seem obvious to us were not attempted or were applied, like the sanitation system of royal palaces in Sumer or Knossos, only to the satisfaction of a dominant minority. Certainly the rate of growth in technology, after the first great outburst of the late neolithic age, was slow, and it was constantly interrupted and set back by the wholesale destruction of cities, with their workshops, and the enslavement and extermination of their artisans and craftsmen. In the vital area of increasing the available energy, there were indeed no gains between the organization of human machines to build pyramids and irrigation systems and the invention of the water mill: that period of roughly three thousand years could show no advance in transforming energy comparable to the increase through food production that took place toward the climax of the neolithic culture, before war had become endemic in urban centers.

The theoretic grounds for despising prescientific technics and protoscience are less sound; for beneath this dismissal is the gratuitous assumption that knowledge not achievable or verifiable by the current methods of science may be treated as nonexistent, even when it has proved far from useless. Our age, at least, should be able to appreciate one of the great virtues of a plodding technology: its very slowness made it possible for a community to assimilate new methods and inventions without disrupting the social order—though even that was not a complete safeguard, for the introduction of coined money in Greece in the seventh century undoubtedly did serious temporary damage. Still, the system operated under a network of human controls and was directed to a variety of other valid purposes besides the advance of science and technology. Even its backwardness has probably been exaggerated because of the disappearance of a mass of early artifacts. In a museum of early American technics such as that at Doylestown, Pennsylvania, you will find a surprising array of resourceful inventions and adaptations, appropriate to a wind-water-and-wood economy, but long discarded and now unremembered. Such isolated historic examples as the process of treating silver to prevent oxidation used by the Damascus weavers, the anesthetic that made possible the wholesale hysterectomies that Herodotus reports, the formula for making the chemically pure iron column found in India, probably point to many other forgotten discoveries and inventions.

Many reasons may be given for the retardation of technics and science: war, slavery, caste segregation, illiteracy, the preservation of trade secrets and mysteries, the monopoly of abstractions like writing and reckoning by a priestly and scribal caste. But perhaps the most important factor has been overlooked, that the pace of production was regulated by a concern for quality. Except in the purely preparatory industries, no product was regarded as complete until it had received a symbolic imprint that made it meaningful and lovable, if not more efficient: sometimes the imprint of form, as in the bowl modeled, the Greeks said, on the breast of Helen; sometimes the extraneous imprint of legend or ornament, as the figures on the fabled shield of Achilles; but tied to human concerns other than the practical or utilitarian. Except for regimented and servile labor, work was never until recent times treated as a purely physical performance; the mere getting of a job done was subordinate to the deliberate enhancement of the operation by colorful ceremonies and playful variations in form that, though they had no practical advantage, added to the human value and significance of the final product.

What Thorstein Veblen used to call the instinct of workmanship likewise slowed down the process, even where no visible aesthetic embellishment or symbolization was called into play. I had my first lesson here long ago, at a furniture factory in High Wycombe which was still turning out the traditional Windsor chairs with rush-bottom seats. The manager pointed out the difference between the machine-made chair and the original craftsman's product. They were superficially the same, but the holes through which the rushes went had been rounded smooth by the craftsman so as not to cut the rushes, while in the factory chair the holes had been left with a sharp edge to cheapen production and hasten the replacement of the more rapidly worn-out seat. This process of reducing quality in order to achieve quantity has become a commonplace in every department of industrial design; indeed, the deliberate lowering of quality is often advocated as the essential mainstay of an expanding economy, though in some departments it has begun to reach a theoretic limit at which the quality is so lowered that the product does not hold together even long enough to be sold.

To dismiss prescientific technology as uninventive is to misapprehend its intentions. But one must admit that a large share of its inventiveness, sometimes a disproportionate share, was in the realm of art: color, pattern, texture, design, symbol. The slowness for which we tend to reproach earlier cultures is canceled out by the long succession of qualitative improvements. Before the seventeenth century this constant humanization was an integral part of mechanization. The empirical knowledge that accompanied technics was of the same order, never entirely cut off from the life-experience and life-purpose of the worker.

This is not to overlook the fact that much work that involved hardship and effort continued to be done, usually under threat of physical punishment or starvation. In the daily tasks of the quarryman, the miner, the brickmaker, the smelter, the sailor, the ditchdigger, there were few aesthetic embellishments: only quantity counted. And it is not by accident that it was from the mine and the arsenal, the least humanized parts of the industrial system, that physicists and technicians after the seventeenth century drew their chief inspiration, for discoveries in ballistics, hydraulics, and mechanics, and for inventions like the steam engine, the railroad, and the hydraulic pump. Apart from this, qualitative standards prevailed everywhere until the nineteenth century: finish, aesthetic richness, meaningful symbolism or ritual were considered more important than quantitative productivity. If anything was to be diminished or sacrificed, it was quantity, not quality. The pervasive concern for quality as related to human interest and capacity served as a built-in control over quantity. Thus the whole tissue of human concerns, religious, social, erotic, aesthetic, constantly played over technics and attached it to purposes and goals that transcended purely quantitative expression.

A satisfactory dynamic equilibrium between quantity and quality is difficult to achieve and maintain; and I would not pretend that any of the higher cultures have ever done so in a fashion that did justice to the needs of its entire population. Like all forms of organic balance, this equilibrium is unstable; and in human communities the regulating system needs constant inspection, appraisal, and readjustment lest it favor one vocation or class. Something like a balance was, however, temporarily achieved in many cities in western Europe by the end of the thirteenth century. Unfortunately, in the fourteenth century that balance was violently upset by the Black Plague, which, on the most careful estimates, wiped out from a third to a half of the entire population. This enormous loss brought about—if I interpret what followed correctly—an immense compensatory effort to repair the quantitative deficiency in both labor and goods: windmills and water mills, mass production of textiles under factory conditions, improvement of sailing ships and canals for economic transport, large-scale operations and ingenious inventions in mining and smelting recorded by Agricola—all these efforts not merely helped repair the loss but redressed a qualitative system that still favored a privileged minority.

We can follow this process of quantification in the growth of capitalist enterprise, with its abstract money economy, in the intensified divison of labor and the increase in labor-saving inventions, many of which, like the fulling mill, the knitting machine, and the domestic clock, long preceded the so-called industrial revolution of the eighteenth century. These efforts all converged toward a quantitative economy, progressively disengaged from qualitative interests and social restrictions: time was money and money was power. The whole process was finally etherealized—that is, transposed into a system of effective abstractions—by the invention of the scientific method. This new method discarded all the human diversions and institutional inhibitions that had hitherto placed limits on purely quantitative procedures. Most people still regard this change as an admirable and beneficent one; and there is little doubt that as long as the new methods of capitalism and physical science were counterbalanced and held in check by a vast array of traditional values, they had a salutary effect in awakening many dormant potentialities. Above all they helped to spread private wealth and so equalize human opportunity otherwise handicapped by scarcity and poverty. But by now the disequilibrium between quantity and quality has become grave: so grave that it may, even without a nuclear holocaust, disintegrate our entire civilization unless more adequate controls can be introduced, and many discarded values and purposes reinstated into the dominant ideology. So it is time to reexamine the basis of the one-sided concern with quantity that we have taken over, all too reverently, from the scientific giants of the seventeenth century. Though their ideological framework is now outmoded in the physical sciences, it lingers on as the working myth and moral imperative of our present age.

I submit that it is the failure to understand the historic relation of quantity to quality which has given a false picture of the radical change that was crystallized by seventeenth-century science and, by this very ideological crystallization, was enormously increased in potency. If I am right, the problems which now confront us cannot be solved in terms of that original ideology. To understand the radical divorce between quantity and quality resulting from the new methodology of the physical sciences, let us consider two figures who arose at the beginning of this period and in so many words define its main characteristics: Francis Bacon and Galileo Galilei. For the sake of brevity I will confine myself largely to their ideas without feeling a necessity to prove how representative or influential they were.

Bacon, in *The Advancement of Learning* and *The New Atlantis,* stands forth not only as the advocate of the experimental method, but as among the first to couple scientific knowledge with technical invention, and to anticipate its manifold effects in multiplying energy, speeding up or arresting natural processes, making possible rapid transportation and instantaneous communication, erecting skyscrapers, in short, controlling and remodeling the whole environment, physical, biological, and social. Bacon's dreams and rational anticipations might have proved sterile were it not that Galileo, Newton, Descartes, and their followers invented an appropriate method for putting them into practice. It is not perhaps an accident that Galileo pursued some of his experimental researches in an arsenal, since war itself had hitherto often been the beneficiary of advances in the physical sciences and technics. At all events, he defined the scientific method as one that paid no attention to the secondary qualities of matter—taste, odor, color, sound. He reduced the significant factors for science to "size, shape, quantity, and motion." The original *Weltbild* of physics had no place for the complex phenomena of organic life: it could handle only dissociated, disorganized, decomposed parts. Even in medical science, exact knowledge was derived chiefly from corpses and skeletons.

Galileo's isolation of physical events from the complex organic manifold in which they actually take place was in itself a decisive and admirable labor-saving invention as long as it kept within its field. By confining attention to measurable quantities and repeatable events, and by confining practice to equally isolated and equally sterilized operations performed in laboratories, the scientists who followed Galileo were able to build up a new world of sensory and mathematical abstractions that corresponded closely enough to purely physical realities to immensely augment their control over mass and motion, time, space, and energy.

This new canon gave priority to physical events, withdrawn from their organic and social matrix; and it threw out, as irrelevant to exact science, as nonscience or nonsense, the rest of human experience, human history, human potentiality. On those terms, the Founders of the Royal Society were entirely justified when they ruled out questions of theology and politics in their discussions. They did this in the name of freedom: but what they called freedom was the privilege, ultimately, of ignoring large and significant areas of human experience.

This was a far more radical displacement of man than that effected by the Copernican revolution. Within the world of science, the new method increased probability; outside that world, it reduced possibility; for more and more human potentiality was limited to mechanical feats that could be formulated by the quantitative method and transposed into the medium of exact technology and reconstituted as a machine or an automaton. The overpowering success of the scientific method within the realm of physical events has long hidden the fact that it has serious limitations. Not least among science's early limitations one must count its forgetfulness of the historic sources of its own subjective bias in favor of quantity and its utter dependence on verbal and mathematical symbols that derive from historic events lying outside the framework of science. Memory and anticipation, feeling and form, function and purpose, are no less real than mass and motion: hence every advance in science and technics, if it is not to carry us farther from reality, must eventually be restored to the context from which the method has, for purely pragmatic purposes, extracted it.

I would stress the deliberately pragmatic nature of the whole quantitative process. Though Galileo had at a very early date defined the methods and the limitations of pure science, the motives behind this change were far from pure, as the words of Bacon and Descartes attest. In his *Discourse on Method,* Descartes clearly states the desirability of using knowledge to gain power: "to discover a practical method, by means of which, knowing the force and action of fire, water, air, the stars, the heavens, and all other bodies surrounding us, as distinctly as we know the various crafts of our artisans, we might also apply them in the same way to all the uses to which they are adapted, and thus render ourselves the lords and possessors of nature." Lords and possessors of nature indeed! That is not a scientific premise: it is the expression of a social program formulated by mind that had far less insight into the symbiotic processes of organic nature than an illiterate neolithic cultivator. But no better ideology could have been framed to serve an expanding economy which was breaking up the complex system of economic protections and moral restraints that had characterized traditional communities. The denial of qualities and the affirmation of quantities went hand in hand, and the less of the first, the more of the second. Multiplication, expansion, automatism, were all written into the new ideological framework that seventeenth-century science had provided; and they applied effectively to the mass production of knowledge as well as the mass production of goods. Both undoubtedly achieved valuable social results. But both lack the necessary human dimensions and human qualifications.

I cannot go into all the implications of this change. But in time the underdimensioned world picture of the physicist supplanted the multidimensional realities of day-to-day experience; and so richly did the new method pay off in practice that it has steadily advanced, even into the realm of the biological and social sciences, whose relative tardiness in development and whose reluctance to dissociate themselves completely from the empiric knowledge already available—and so valuable—in medicine, agriculture, and politics had long permitted them to draw surreptitiously upon experiences and insights that do not submit to scientific processing. The general effect of regarding the purely quantitative methods of seventeenth-century science as the highest type of order was to dismiss qualitative criteria and disregard the more organic and stabilizing processes, thus ignoring the essential human need for continuity, for accumulation, for maintaining variety, for rational assimilation and integration, for the interrelation of parts into ever more complex wholes.

Now the conquest of nature is not, from the standpoint of the somewhat more adequate biological and social knowledge we possess today, the supreme command or the unalloyed blessing that our scientific technology naively takes it to be. In terms of lording it over nature, the engineer who uses a bulldozer to root out trees, to remove the topsoil, and level hills into flat tracts that can be covered with asphalt and planted with uniform rows of mass-produced houses represents a higher form of technology than that of an architect who respects the contours of the hill, keeps every possible tree and bush, replenishes the topsoil, designs gardens for aesthetic delight, minimizes the paved-road system, and provides houses individualized in plan to meet the requirements of their occupants, in numbers that do not lessen the habitability of the site. But because quantity remains our main criterion of success, particularly if quantitative effort produces monetary gain, the bulldozing engineer has become the paragon of our civilization and is everywhere wiping out organic variety and human quality. The bulldozer's method and the human animus behind it were latent in the formulation of the scientific method in the seventeenth century, albeit Bacon was a courtier who loved gardens and spent fabulous sums on entertainment, and Galileo was a humanist of distinction, at home in the world of literature and philosophy.

I intend no diatribe against either science or technics, so long as they remain subordinate to organic functions and human purposes. I am merely trying to explain how their immense human benefits were curtailed by a one-sided overemphasis on quantity, and the exercise of a one-sided control over both nature and man, who speedily became the victim of his own favored method.

By isolating the scientifically observable fact from its social context and detaching the disciplined scientific intelligence from the whole human personality, science lent itself to the processing of accurate knowledge, and this knowledge has been of immense practical use. But science did this by disqualifying man himself from directing the process, since the very symbolic and social concerns that had proved effective for controlling quantity in the empirical, prescientific period were dismissed as having no status within the world of science and technics, where only quantity counted. Human purpose could not be altogether eliminated from this new system; but it came back in disguise, as a thirst for limitless power over nature—the power to create human equivalents of natural organisms, in the form of machines or synthetic quasi-organisms; and the power potentially to destroy all organisms now so amply satisfied by the weapons made possible by nuclear physics. The end product of this choice is now in sight: power, thus methodically pursued, without natural limits or human goals, is rapidly becoming paranoid for lack of the very dimensions of life that it rigidly excluded: historical knowledge, political experience, human sensibility, parental feeling, the habits of loving and nurturing and preserving the forms and expressions of life.

What is most threatening in this obsolete system's lack of qualitative reference derives from its very success in throwing off historic restraints; scientific and technical advance has become automatic and compulsive: not merely compulsive, but, humanly speaking, manic —although the neutral word for this aberration is dynamic, now often used as what Robert Frost calls a praise word, as if the static aspects of reality were to be dismissed as unobservable or unreal. Both Norbert Wiener and myself have used Goethe's apt fable about the Sorcerer's Apprentice to illustrate our present plight in every field, not least in the proliferation of scientific knowledge. Goethe tells how the magician's young assistant was left alone to clean up the workshop; having acquired the master's formula for making the pails and brooms work without human hands, the lazy fellow invoked that spell, only to find that his mechanical servants worked so swiftly and automatically that he was almost drowned in a flood of water because he had never mastered the magician's formula for slowing them down or turning them off.

This is the plight of our present economy. Standardization, mass production, quantification, have overwhelmed us. Our favored ideology, which has proved so successful in dominating the physical world and building up a kind of secondary community of mechanisms, has no built-in brakes or regulators and no place for limiting purposes or goals that lie outside the system. In the fabulous brave new world that is already visible, electronic computers will play chess with computers, machines will reproduce other machines without human aid, cybernetic brains will match wits—and game theories—with other cybernetic brains equally unconcerned with the history and destiny of man, until finally rocket systems lacking any instinct of self-preservation will exchange nuclear destruction with other rocket systems. As the economy increases in power and automatic perfection, man will be literally wiped off the map.

If this diagnosis is sound, the remedy will not be simple or swift; but we can at least point to where it will take place, if not precisely how it will come about.

Though the present situation is in no small measure the consequence of the quantitative methodology and the concealed social imperatives of the seventeenth century, when we look for a way of reinstating qualitative controls we must first take account of significant changes that have taken place in science itself. Exact science is no longer confined to the physical world; it has moralized and disciplined the search for sharable knowledge in every department. Orderly method, predictability, communicability, cooperative verification, are significant values, and produce qualitative benefits, though these benefits were not altogether lacking in the older empirical tradition. But the properties of the physical world only partly suffice for organisms. The biologists of the nineteenth century, centering around Darwin, introduced new categories and new properties, peculiar to organisms: autonomous patterns of behavior, life histories, and symbiotic associations that exist in elementary particles or charges only as latencies or potentialities. The life cycle, the reproductive cycle, organic balance and autonomy, form-maintaining and goal-seeking processes—all these are qualitative properties; indeed, the very concept of organism implies the existence of built-in regulators, external and internal, to control the intake and output of energy, to limit growth, to maintain balance, to discriminate qualities, to pursue goals.

There has been a serious cultural lag in assimilating these new organic and ecologic insights. In our generation, indeed, there has been a definite retrogression; for, in an attempt to emulate the magical successes of nuclear physics, many biologists have centered their attention upon processes or isolated fragments of organic systems, like the genes, that can be brought under physico-chemical control, and modified at will.

Quality in control of quantity is probably the great lesson of all biological evolution, so far as that reveals the progressive emergence of nervous systems capable of dealing with qualitative choices and discriminations. The secondary properties of taste and odor that Galileo properly discarded as irrelevant in a study of mass and motion are essential to survival in the organic world; and at higher levels, which birds, dogs, and many other creatures share, other qualities, like companionship, sympathy, love, and beauty, may be equally vital to nutrition and organic vigor and psychological balance. To introduce these qualities at too low a level, as physicists did when they misinterpreted planetary motion as circular because of the circle's supposed perfection, or as chemists did when they spoke of chemical affinities in terms of human attraction, is bad science; but it is equally bad science to reject these qualities at the biological or human level when they are essential to an adequate description of organic functions and human purposes. This lesson, so remote from the world of physics, has still to be absorbed.

Similarly, many qualitative improvements have been taking place in technics: particularly in departments, such as electrical engineering and electronics, where the impact of scientific thought has been most direct. An automatic telephone system is a vast qualitative improvement in human terms, in that it releases thousands of human operatives from the slavery of the switchboard; and even the quantitative extension of the telephone has a qualitative significance, in that it potentially makes all the participators in the system members of a humanly richer community, if they speak the same language. Though such human communication is unfortunately reaching a limit for lack of phonetic research and linguistic invention comparable to that in electronics, similar qualitative changes have been going on in other departments. The current aesthetic revulsion against the vulgar deformities of Detroit's favorite motorcars, with their absurd wings, their inflated dimensions, their squat forms, has already effected a curb on the productive fecundity of the American motorcar, and favored the production of smaller cars of more refined design. The very surfeit of quantity in many departments may in time beget a compensatory passion for quality.

We are now back at the point where we started—the control of quantity by quality. If I am right, to reestablish this control demands an ideological transformation fully as bold as that which was effected by the great scientific minds of the seventeenth century when they turned their backs upon the First Causes and Final Ends that had dominated medieval theology and its residues of Aristotelian science. These minds produced a new world in which physical events had primacy by definition: a world favorable to quantity and largely indifferent to quality. We have now just the opposite task. To restore balance, we must address ourselves primarily to organisms and personalities and regard their isolatable physical aspects as secondary to the formative processes, the purposeful associations, and the meaningful goals by which communities as human communities live and flourish. In this new ideology, quantity will not reign supreme: it must always be modified and justified by quality. Our object is no longer the one-sided domination of nature, but the creation of sympathetic associations and cooperations favorable to life and to vivid intercourse with nature on many levels besides the physical one. To this end, we will seek not the maximum quantity of energy or the maximum power of external control, but the right quantity of the right quality at the right time and the right place for the right purpose. It is in this context that whole men, rather than perfect machines, mutilated organisms and underdimensioned men, will flourish.

RESOURCE QUALITY AND VALUE OF THE LANDSCAPE   *D. B. Luten*

When we speak of the "quality" of a natural resource, two aspects may come to mind. The first is well symbolized by water. The "quality" of water is often an issue in the management of water, but it can usually be measured in terms of the quantity of impurities. While the qualitative epithets "good" or "bad" may be applied to it, they always lead to the queries, "How good?" "How bad?" And the answers must be made in quantitative terms. In sharp contrast is the second aspect: the "quality" of a sparrow compared to that of a robin.

Two theses are examined in this essay: the first pertains to divisions and boundaries, matters necessarily associated with the latter sort of quality. It is discussed partly because the topic of resource quality pervades this book, but also because it is a reasonable preliminary to the second argument, which is an effort to justify the landscape itself as a natural resource.

Any discipline has divisions, which must be arranged. An arrangement of divisions is properly called a "taxonomy" (Gr. *nemein,* to arrange; *taxis,* a division), but the word is commonly used only in biology. The first part of this essay undertakes to broaden the idea by referring to divisions of the material objects in the landscape.

Divisions are usually separated by boundaries, and boundaries are important because that is where qualities change. It is at the boundary that a robin might become a sparrow. But both robins and sparrows are birds, a generic quality, and can be counted on a common basis: quantity transcends boundaries.

Boundaries which separate domains of differing qualities have their own attributes. In nature, some boundaries are clear-cut; others are vague. The earliest recognition that species exist came perhaps for animals and plants. Birds are birds and do not grade imperceptibly into mammals or plants, or from crows into crocus, from dogs to dogwood. Wood's book, *How to Tell the Birds from the Flowers*,[1] is not serious, at least not very much so. Most of the individuals of a biological species closely resemble the average individual, but one species resembles a second more closely than it does a third. So it is helpful, it exposes a pattern, it appeals to our instinctive search for order, to arrange these divisions in an orderly manner.

In chemistry, the search for divisions was not fruitful in its early stages: earth, air, fire, water. The boundaries were not satisfactory; the individual examples were not tightly clustered around a type. In the eighteenth and nineteenth centuries the situation changed dramatically. Now it is clear that chemical substances are diverse, have specific qualities, and, almost inescapably, are classed as species. The species are arranged, in one fashion or another, according as we discern their relations. In organic chemistry the arrangement is the structural theory. So we have a taxonomy of chemistry because we have an arrangement of divisions. It is not the *purpose* of chemistry to prepare a taxonomy any more than it is the purpose of biology. But in each science, at an early stage in the search for order, the development of a taxonomy might have seemed a primary purpose.

Since chemistry is concerned with the transformation of matter,[2] it can be concerned only with understanding the rates and equilibria of chemical transformations. The structural theory of organic chemistry is only one of a number of accessories to that understanding. It is the greatest of these, and a highly complex and successful abstraction, but it is only an accessory. The purpose of the structural theory is to assist in the understanding, to reveal the patterns of order in the transformations of matter.

Chemical taxonomy shows what species may exist, and shows good and bad paths for transformations, for travel, from one species to another. It speaks of these paths in terms of equilibrium and rate. It can and often does speak authoritatively of equilibrium, a timeless matter. In contrast, when chemists speak of rate they cannot be strongly predictive and must bring in a new entity, the environment. At an early stage, it seemed a reasonable undertaking to try to change lead into gold; later the possibility was denied; still later, chemistry suggested that a path might exist, but the equilibrium seemed unfavor-

---

[1] Robert W. Wood, *How to Tell the Birds from the Flowers* (San Francisco, 1907).

[2] Some of the material in these pargaraphs is from D. B. Luten, "On Chemistry and Taxonomy—Both Biological and Chemical," *Lloydia*, Vol. 27, No. 2 (June, 1964), p. 135.

able. To my knowledge, the environment for encouraging this transformation has not been defined.

Again, chemistry says that you cannot easily change methane to carbon dioxide except in an environment of air. Measurement has shown that the equilibrium is favorable and, in a proper environment of temperature and concentration, the transformation is easy, sometimes too easy. The taxonomy of chemistry, then, guides us quickly to the conclusion that ethane and a host of other chemical species will behave in a similar fashion.

Chemists long ago learned the importance of environment and pretty well how to define it, having in mind the entities of reaction partner, temperature, medium, and concentration. Environment, though, is not a part of the taxonomy. We can envisage a boundary —perhaps "envelope" is a better word—which divides what is in the environment of a chemical transformation from what is not. But the only division here is between "in" and "out." It is not a division of one part from another. And so we do not have a taxonomy of chemical transformations. If we had, might we not have a full understanding, a synthesis, of the transformation of matter? But if we have not solved this problem in chemistry, it is hardly surprising that it remains an obstacle in other fields.

The divisions of chemistry have been arranged to show relationships, and these relationships are useful in predicting the transformations of matter. What of the divisions themselves, of the boundaries between them? They are not arbitrary; chemical substances do not intergrade continuously in all conceivable directions. Chemical substances occur in species and in groups of species just as biological organisms do. The speciation, in concept, is utterly sharp. And it is *objective*. There are a number of instruments which will, when traversing some quantitative property of a group of substances, draw a curve closely representing this perhaps infinitely sharp speciation. Mass spectrographs and gas-liquid chromatographs are two such instruments. The curves are illustrated in figure 1. The quantity measured in the one case is essentially molecular weight, the mass of an individual of the species; in the other it is the volatility of an individual species. The curve shows that for most values of volatility *nothing*

Fig. 1

exists, but that at certain values a species is found. Two species are usually separated by emptiness. You could hardly find a foundation in this world which would sponsor a search for an intermediate between methane and ethane. The taxonomy says clearly that there should be none.

No a priori reason for such speciation has been brought to my attention, and I believe none exists. No a priori reason exists to say that elements should not intergrade continuously, that their compounds should not vary continuously in composition and properties. So deeply imbued with the notion of sharp speciation are we that it is hard for us to envision the problems early chemists encountered in establishing that such is the nature of matter.

In biology, matters are different. The taxonomy does not predict transformations very well. Underlying biological taxonomy is again the idea of relatedness, but in a one-way temporal scheme, the idea of a ramifying evolution. Biological taxonomy displays the patterns of order of the evolutionary process. As in chemistry, it does not include the environment directly. If we could manage to include the environment in the taxonomy, would we have a more predictive taxonomy? Could we know what to expect of the future?

In the living world, just as in chemistry, speciation occurs. Robins are not sparrows and do not intergrade into sparrows. They are thrushes, but thrushes other than robins exist. Species exist and groups of species exist, and patterns of order among them can be discerned, and so we can establish an arrangement of divisions which aids the mind in encompassing what is known. But the taxonomy of biology will not help us to transform robins into sparrows. Instead, it will tell us something of when and how some common ancestor came to be transformed into both robins and sparrows.

Speciation in biology is less sharp than in chemistry. Individuals do deviate from the typical, but the distinctions even between closely related species are usually objective and substantial (fig. 2). Thus in both chemical and biological taxonomy, sharp boundaries exist between species and, without doubt, quality changes at the boundary.

Another example that comes to mind, perhaps because pedologists

Fig. 2

have done so much classifying, is soil. The environment has been directly involved in the arrangement of divisions, and its influences, climate, parent material, and age are easily seen in the taxonomy. This arrangement of divisions seems to aid our understanding of the ramifying evolution of soils and the prediction of transformations. But its precision is not high and perhaps this stems from the nature of the species of soils. Species certainly exist, and boundaries separate them, both on the land and in the taxonomy. But speciation is no longer sharp, and the variation of soil quality as we alter some quantity, whether it be miles across the earth's surface or a parameter such as acidity, is rarely sharp (fig. 3). This stems from the absence of

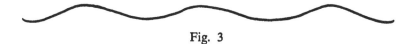

Fig. 3

sharp boundaries in the environmental factors which generated the soils, but also from the absence of factors tending to eliminate soils which intergrade, which tend to deviate from a type.

Thus, in three instances chosen from a discontinuity of phenomena, a continuum in the nature of the boundary seems to arise. Schultz [3] has suggested that this should be related to the complexity of the systems in which the manifestations are found. When a system may be defined with but a few parameters, boundaries should be sharp; when many are needed, discontinuities become weak.

In three other examples, chosen from a host of closely related sorts of arrangements of division, no real discontinuities or even rapid changes of quality occur at or near the borders, but only changes in quantity. But in the first of this group, changes in quality certainly occur within the arrangement of divisions. Where is the border between hard and soft water, or between small cities and large? Still, hard and soft water do have differences in quality, and small and large cities as well. But the borders now are set arbitrarily and a continuous range of quality closely related to a change in quantity must be expected. In the second example, the arrangement of divisions may be an administrative device. What change occurs at the border between the 16th and 17th assembly districts? None; but an astute observer can itemize many differences in quality between the two districts, each taken as a whole. However, this is largely descriptive; pattern, order, is not disclosed by the arrangement. The third example in

---

[3] A. M. Schultz, personal communication.

this group is simply a pigeon-holing device, employed for mnemonic reasons.

The common attribute of all these examples is that the boundary has become needle-sharp, as it was for the boundaries of chemical species, but the curve is inverted (fig. 4). Here, in place of sharply

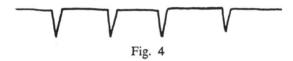

Fig. 4

defined species separated by emptiness and discontinuous changes in quality, we have a fully occupied region with arbitrary boundaries at which no sharp changes in quality occur. Quality now varies continuously and more or less uniformly.

We must refer to the boundaries of chemical species as natural boundaries and to these last as artificial. From all of the above we conclude that *quantity transcends boundaries,* and *quality changes at boundaries.* But, also, inevitably we shall say that *the better the boundary, the more quality changes at it.*

The foregoing was stimulated in large part by the title of the continuing discussion of which this book is the record: "Quality and Quantity in the Management of Natural Resources." It is introductory also to my second thesis, that landscape, especially the wilderness landscape, can be demonstrated objectively to be a natural resource.

The "envelopes" of systems, apart from being very different from taxonomic boundaries, may be quite diverse among themselves. But their diversity reflects the attributes of the systems they envelop. Thus an envelope surrounding a system may be material and may separate the system physically from the surroundings, or it may be imaginary and serve only to focus our attention on what lies within it, to focus our attention on the system. Chemists, especially, have created but also imagined systems in which no energy passes through the envelope. These are called "adiabatic" (Gr. *diabatein,* to pass through), but there is no particular reason to stress energy as the entity which is not to pass through.

Examination of an adiabatic system usually begins with a disturbance which displaces it from equilibrium, followed by observation of the processes of its return to equilibrium. This can be reasonably approximated in the laboratory, less so in the field, where we more often encounter steady-state systems, whose adiabaticity is impaired by a flow through them. Most commonly, perhaps, this is a flow of energy, and we are familiar with biological systems into which energy flows as sunlight, and from which it leaves as infrared radiation. Or matter may enter as a stream of water and leave in much the same condition. If such flows are limited in number and reasonably well understood, they may be no handicap. Sometimes they may even be helpful. Take, for example, what chemical engineers call a continuous stirred tank reactor. This is a vigorously stirred vessel into which reactants flow continuously and from which an equal stream discharges. The chemical reaction kinetics of the processes occurring in the tank are often simpler than those of an isolated mass-adiabatic vessel, and especially of an isolated mass-energy-adiabatic vessel.

So we may have material or conceptual system envelopes of varying adiabaticity surrounding a contiguous region in space. But there is no reason why a system must be contiguous. Instead of having sunlight pass through the envelope, the system may reasonably be extended to include that part of the sun which is generating the incoming light and that part of the cosmos which is receiving the emitted infrared radiation.

This alternative idea of an envelope is that it should include not only the object of interest but also its significant environment. Entities, even though remote in ordinary space, should be included if they influence the object of interest. Entities, even though neighboring, to which it is indifferent may and should, in the interests of simplicity, be excluded.

The envelope for this system may become a curious sort of surface: it may be discontinuous, but one side of it will be "in" and the other "out" of the system. However, it is also selective; so some items of matter, or of energy, which at first seem "in" are actually "out." This is because they are without influence on the object of interest. Thus sunlight may be "in" while radio waves in the same region are "out" of the envelope. The crucial test is whether such items are part of the significant environment. Perhaps we can go so far as to ask whether, in some resource discussions, economic matters might be "in," but aesthetic factors "out" of the system.

Having played hob with all reasonable ideas of arrangements of a system, let me now stipulate one more outrageous condition: imagine that the significant parts of the environment are located in a pie-shaped region centered on the object of interest at distances proportional to their influences. Thus we have an object of interest at the center of a system and arrayed around it most closely are the least influential parts of its environment; at a greater but still modest distance are the most influential parts of the environment. The object is at the center; the greatest influences are at the rim.

The object of interest is taken to be biological, an individual, the members of a species in a region, or all the members of a species. We shall call it a biosystem, not an ecosystem. Were we to include all the members of a community of different sorts of organisms we might have an ecosystem, but this is a thornier notion and I should like to shy away from it, except for a passing note later on.

I have chosen this kind of arrangement in order to consider what happens to the system with the passage of time, and this arrangement provides an explicit analogy to a simple physical system. In one device used by physicists in accelerating nuclear particles, the Bevatron, electrons are guided on a circular course by an array of magnets and electrical fields located on that course. The environment of an electron, or a small group of electrons, is the magnetic and electrical fields within which it finds itself. They constrain it, guide it; they dominate it. At any instant, the environment of an electron is the field of the nearest magnet and the immediate electrical field, not the fields of the magnets farther along. In a later instant, the field of a subsequent magnet and the electrical field in that neighborhood will be the environment. The environment is pie-shaped at any one moment; over a period of time it resembles a cylinder and, with the Bevatron, it is a bent cylinder. When I speak of the "bent cylinder" as the system moves in time, I am trying to suggest an environment which may be changing, an environment which disturbs, which modifies, the electron's path, not that time itself is curved.

In the Bevatron, some members of a group of electrons, because of secondary forces, interactions, perturbations, disturbances, stray from the central path essential for success. As they do, focusing forces in the fields tend to bring them back on the paths desired for them by the designers of the equipment.

In the biosystem, the environment which is significant to the organism is, by that token, the part of the environment which constrains the organism, which directs it on its course through time. We can view the organism as constrained within this pie-shaped region. What lies outside the rim has no influence; within the cylinder, what lies behind or ahead is a past or future environment. As a corollary part of the analogy, picture the well-adapted organism as being close to the center of the pie, secure in its adaptation and free, momentarily at least, of pressure from the environment. But when an individual strays from the center, because of perturbations current or recent, it is subjected to constraints, to pressures, to forces. The more the straying, the greater the forces. The strongest forces come into play only with increased straying. The most positive constraints are the most remote, not the most immediate, parts of an environment so arranged. An individual which strays too far will completely escape the system and disappear.

Since biological speciation is not perfect, individuals do tend to stray from the course of perfect adaptation, of conformity. In the biological environment, I think I see some direct focusing forces which tend to bring individuals back into line. Perhaps more commonly, however, individuals which tend to stray, whether geographically, or physiologically, tend simply to disappear, to be wiped out. But over long periods of time, surely a focusing effect exists. In each generation, the offspring of mild deviants contain more deviant, but also less deviant, individuals. The more deviant are eliminated in favor of the less, and the less deviant provide the next generation. We must keep in mind that the environment is not immutable. An organism deviant today may, without changing, be conforming tomorrow because the environment has altered.

At this point I want to try to identify "ecosystem" in terms of the last few paragraphs. An ecosystem would appear to be an association of biosystems such that each organism is a part of the environment of every other organism of the ecosystem. The concept of space devised in earlier paragraphs to describe a biosystem is distorted enough; the intertangling consequent on the extension to the ecosystem is much more involved. Perhaps, indeed, so much that the image becomes fruitless.

Returning to the simpler idea, picture the environment as a guide which keeps the organism on course, as it were, but as a guide which itself may vary with time. If the environment were to remain unchanged, would the biota remain unchanged? Perhaps so, but, because much of the environment is itself biological (the biosystem is a part of an ecosystem and entangled with it), the chances of evolution coming to an end are remote.

In fact, though, the environment does change, and as it does, it guides the organism at its nucleus. The electron's nature is to go in a straight line; the Bevatron's magnetic and electrical fields cause it to veer and constrain it into a circular path. When the environment changes, the organism rubs against one side and rides free on the other. On one side the environment is constraining; on the other, beckoning. If the organism generates deviants that veer toward the beckoning opportunity, it adapts; it evolves, generation by generation; it is viable. If the environment changes too rapidly for the adaptability of the organism, the organism escapes the environment altogether and becomes extinct. While the environment persists, the biosystem vanishes with the escaped organism, for the organism is as much the essence of the system as is the environment.

How and why does the environment change? For my immediate purposes, ignore the changes in the inorganic world. They are by no means negligible: glacial periods, subsidences, new surfaces, chemical changes, all these have affected innumerable biosystems drastically and rapidly, and have influenced evolution, but that is not the immediate issue. Let us consider only the changes due to the activity of our central organism or of other organisms in its environment.

The most obvious way in which an organism changes its environment is by eating it up, but many others come to mind: by shading out one's offspring, by changing the pH of the soil, by building a next, by claiming a territory, by developing a sod, by producing smog. Eating up the environment might have been dramatically rapid when this world's first freely reproducing organism went to work on the ocean of broth which nurtured it.[4] There may well have been oscillations in the nature of that environment so spectacular, so much beyond adaptive capabilities, that the creation of life itself had to be repeated. Still, at some point, and long ago, a relatively stable accommodation was reached. Since then, alterations due to activities of the biota have been mostly slow. As one organism changed, owing to the constraints and opportunities of the environment (the rubbing and the beckoning), it, being a part of the environment of its neighbors, caused changes in the neighbors' constraints and opportunities. If the neighbor could adapt and respond to opportunity, it thrived; otherwise it vanished.

Next, consider two limiting cases: in the first, the organism has no effect on the environment. Its associates change its environment. It is, perhaps, in no biosystem but its own. That is, it has no significant influence on any other organism. This is the predicament of the rare organism, which must conform and adapt, as circumstance dictates, or die. Its survival depends on the rate and manner in which the environment shifts, and on the organism's ability to adapt, to stay somewhere near the center of its biosystem. The species need not remain invariant. The important test is whether it has descendants, whether of the same species or of derived ones.

The record suggests that not many organisms have met this test: the limbs of the phylogenetic tree are intricately branched. Few primitive

---

[4] See Garrett Hardin, "A Second Sermon on the Mount," *Perspectives in Biology and Medicine*, Vol. 6, No. 3 (Spring, 1963), pp. 366–371.

organisms have left descendants, but those which have succeeded have left many and diverse descendants. The environment of most organisms of the past has at some time veered so sharply that the organism could not stay within it and was accordingly wiped out. But while this was happening, the environments of other organisms were opening up and ceasing to constrain. And the opportunity for diversification, for individuality, was expressed in ramification.[5]

In my second limiting case, the organism is completely in control of its environment in the sense that all influences on the environment come from the organism itself, not from any independent source. It influences, often dominates, the environments of its associates while they have no influence on, are not part of, its biosystem. Here is an organism which can do no wrong. No matter how it veers, its environment veers with it and it continues unimpaired on its own course.

This is a curious situation, for we must exclude any idea that the course is guided. This leaves us in a predicament: if the course were utterly random, without constraints of any kind, it would also be discontinuous, here this instant, there, perhaps remotely there, in the next instant. It becomes a rather useless sort of limiting case unless we impose some restraints on how the course may vary. Yet, if the course is not to vary without restraint, it must be guided in some fashion.

Three kinds of guidance come to mind: foresight, inertial, and by a fixed reference point.

Much has been said of foresight, of purpose, of setting out for a target, but, by general admission, hindsight is sharper than foresight. Nonetheless, all life is concerned for the future, so all life has some vision of the future, and a notion of what it will be like, and prepares for anticipated events. Much anticipation is genetic, but not all. Foresight serves to avoid a serious rubbing against the environment, and leads to a veering when it is judged the environment is about to veer. "Go to the ant, thou sluggard." It is as if the organism were probing ahead for turns in the road and had learned from the past to prepare for such turns, to maintain a conservative course even when constraints appear remote.

Currently, we hear much discussion about whether humanity, while apparently in full control of its environment, is in fact approaching a corner too sharp to be rounded, or even creating a corner too sharp to be rounded. An easy example is the human use of energy. For

---

[5] See Garrett Hardin, *Nature and Man's Fate* (New York, 1959), p. 75 and chapter 12.

millennia we have been great users of energy, converting chiefly from chemical latent energy to heat. In recent decades our rate of energy consumption has rapidly accelerated, and now we vacillate between the belief that fuels are limitless and the fear that they are nearly exhausted. One appraisal [6] has suggested a 7 per cent increase in energy use each year. This is too high, perhaps twice too high, but it is a convenient number because it comes to a thousandfold per century. If 3.5 per cent a year is more accurate, think of a thousandfold increase every two centuries.

Men are essentially in control of this part of the environment. Perhaps we can continue on the exponential growth curve, in our use of energy. At this moment the energy resource seems to impose no limit, and the amount we use today is still small, about $\frac{1}{30,000}$ of the sun's input to the earth. But a 7 per cent increase for the next 150 years would amount to a 30,000-fold increase in human use, and would cause it to equal the sun's input to the earth. It is hard to imagine what anyone would want with so much energy until near the end of that interval, but then a need becomes terribly clear.[7] With double the amount of energy released and double the amount to be radiated from the earth's surface, the average temperature would be 100°F higher than now. I envision most of that energy being used for air conditioning. Would such a circumstance be an unmanageably sharp turn in the road, an intolerable constraint by the environment, an unescapable trap? The answer is that our foresight should be adequate to guide us away from this particular hazard, but our prevision of other hazards may be less sharp.

Human foresight can do wonders with such problems. That it can serve as a guide for the environment must be denied. That would be possible only if a target, a clear human purpose, existed. Who is wise enough and respected enough to define such a purpose?

The second sort of guidance, the inertial, may be criticized on the score that it defines the future in terms of the past. But if the future is to endure, it must look to the past for guidance. A host of inertial guides may be imagined: rate of genetic variation; dominance of the group over the individual, that is, convention; rate of communication within the system—how quickly is the information that the organism

---

[6] Anonymous, "Fusion Power—Future Necessity," *Chemical and Engineering News,* December 24, 1956, p. 6290.

[7] See D. B. Luten, "Metropolis in Flood," California Society American Institute of Park Executives, Proceedings of the Twelfth Annual Conference, January 13, 1961, p. 7; D. B. Luten, "Parks and People, *Landscape,* Vol. 12, No. 2 (Winter, 1962–1963), pp. 3–7; J. H. Fremlin, "How Many People Can the World Support?," *New Scientist,* Vol. 26 (October 29, 1964), pp. 285–287.

has changed its course communicated to the environment so that it can veer to match? Alternatively, how much of the information thus provided reaches the environment so as to modify it? These guides encompass memory, tradition, conservatism, be it learned or genetic.

An organism dominating its environment, its biosystem, and controlled by inertial guides may be realistically imagined. Most such systems, I suspect, are associated with population outbursts. However, do not include in this category systems whose transiently unconstraining environments permit an outburst of population, for they do not modify the environment in their own favor. In contrast, some examples of marginal control of environment are the tree which modifies the soil in which it grows to make it better for its requirements; the bird which builds a nest and claims a territory.

An organism in this category would dominate the environment of its associates without being affected by them. No matter how much it might veer, its environment would veer with it. Fixed reference points and foresight being excluded for the moment, this biosystem has no Polaris, no fixed star for the helmsman, no clear target. At best, its guides would be memories of the past, the ghosts of history peering into the heavens for a Polaris, for a purpose. Instead of these, it would have inertial guides: the inertia of innovation (How fast could technology solve the problems it has created?), the inertia of delay in communication (but both communication and change in cities seem ever to accelerate). What portion of advocacy of a new course would be heeded? What conformity would a society demand? How fast could traditions change? To what degree could the repositories of traditions be shielded from the pressures of a society?

How fast could such a course change? Would it be an exponential spiral? Genetic adaptability would eventually become an external constraint to limit the fate of such a system. Exactly how is unclear. Mankind appears to be gaining so much control over his environment that he is approaching this condition. But there is no guarantee of perpetual control. Inevitably, control will pass back and forth between organism and environment, and disaster beckons at each transfer. The image, in my model, is one of getting into too tight a turn, of losing control and being abruptly wiped out.

The third kind of guidance is the reference point. If a system in which the organism controls the environment is to maintain a stable course, if it is not to veer uncontrolled, if it is to be homeostatic, it must have a reference device, an immutable Polaris.

A familiar instance of homeostasis is homeothermy. A simple control system such as an ordinary thermostat has a reference temperature device, perhaps a bimetallic element, perhaps a mercury thermometer with electrodes in it. We have these familiar devices in our laboratories, in our homes. We also have something of the sort in our bodies. A competent textbook on physiology [8] says a great deal about how information is relayed in the body to modify the output of heat, but not a word is said about the reference device except that it is probably located in the hypothalamus. How does the body know that 98.6°F is the normal temperature? How do three billion human bodies each know this same thing? Only because a reference device is there to keep body temperature constant, on course, in spite of drastic changes in the environmental temperature, until the verge of death. It is hard to imagine what the device might be other than a phase equilibration (resembling the melting of ice) or a group of opposed chemical reactions. Its nature is unimportant just now; its existence is important.

We should not expect to find reference devices in biosystems, for they are dependent on their environments and have no use for such devices. We might consider the tangled web of biosystems which make up an ecosystem, for it has a measure of stability, and could have something similar to a set of opposed reactions. Whether such a control depends unequivocally on a reference point or is really inertial is not clear, but neither is it important at the moment. While a reference device may be small, this is not necessarily true. Conceivably it might have to be very large. How large is Polaris?

I have cited three influences which may serve to keep an organism on a course, despite a growing dominance over its environment. One of these is inertial, the rate at which the biosystem can change. Another is foresight, a probing into the nature of a future environment with responsive modification of the course. A third is the existence of a specific reference mechanism, be it Polaris, a bimetallic regulator, a mixture of ice and water, or what not.

If human society should become independent of its environment, what might its course be? Is there any limit to the wandering of the system, or in the rate at which it veers? [9] The answer has been sought in an outpouring of writing during the past two decades, in the controversy between "Enough and to Spare" and "Our Plundered Planet." The answer seems to be that there is a limit but heaven only knows where.

---

[8] A. C. Guyton, *Textbook of Medical Physiology* (Philadelphia, 1956), pp. 950–968.

[9] See Lewis Mumford, "The Human Prospect," *The Role of the Region* (An Institute on Planning for the North Central Valley, University of California, Davis Campus, 1962), pp. 29–39.

Should we suppress mankind's dominance so that we become subservient to our environment? No; that is impractical, unrealistic, unachievable.

Does mankind have foresight? Yes, a great deal more than we can perceive in other organisms. Does it keep us on constant course? No; on the contrary, foresight serves only to increase human adaptability, to accommodate the tendency toward exponential spiraling of our environment, itself resulting from the diminishing of inertial controls. Does the human biosystem have a built-in reference device? None that is obvious to many of us.

I can see only one possibility for such a reference device: the traditional environment of mankind, the native landscape, the natural scene, the wilderness. This environment, to which we were subservient for a million-odd years, is now greatly altered, and survives unchanged only in a few remote or forbidding retreats. (Does enough of our traditional environment remain to provide an adequate reference device?) The manner of operation of such a reference device is not clear. Whether it is a promising one must be argued, but even if no answer is forthcoming, it is still the best hope until a better one is suggested.

While it is probably inborn for men to struggle against the dominance of nature, it is probably inborn also to recognize this dominance and submit to it. The more people remove themselves from the natural scene as they accede to the pull of the cities, the more we find them seeking, as opportunity permits, a return to the open land. Olmsted's words of a century ago citing the need expressed by government leaders for contact with the natural scene are worth remembering.[10] As urbanization grows, the descendants of men who saw no good in the natural scene now turn back increasingly toward it. Is the oscillation between city and countryside perhaps a manifestation of the need for guidance by the unique reference device?

Concerned with such matters these days, we find two sorts of people. One group can be categorized as those who understand immediately and are awakened by and sympathetic to Sigurd Olson's phrase, "The Singing Wilderness." For the second group, the wilderness is only land awaiting development, and the symbolic phrase is "a howling wilderness." Who, in terms not of the immediate but of the longer-range wandering of the human-dominated environment, is the deviant and who the conformist? Against whom does the human environment rub and whom does it beckon?

---

[10] See Laura Ward Roper, "A Preliminary Report (1865) by Frederick Law Olmsted on the Yosemite Valley and the Mariposa Big Trees," *Landscape Architecture*, Vol. 43, No. 1 (October, 1952), pp. 12–25.

It must become clear to all who are concerned for the future that the natural scene is the only conceivable reference point. Further, it must be retained, even though we cannot understand how it could work, until the day we are able to prove that it cannot work.

How much of the natural scene is necessary depends on such matters as the amplification of information it provides and on how much it is disturbed in the process of providing that information. The amount needed might be very little, perhaps a forty-acre natural prairie; the bimetallic unit needed to control the temperature of a house is very small. Still, prudence would suggest maximum retention until the needed level is clearly established. Common sense would suggest that the more abundant the natural scene, the more effective will be its control and the less erratic the veering and yawing of the environment under the guidance of man.

How perfect, how unimpaired must the landscape be to be effective? No good answer is at hand. Landscape of any quality exerts some sort of control, whether of an inertial nature or as a reference point, but, because only an unimpaired wilderness can be independent of the human influence, only it can be a dependable reference point.

I have not tried to say that wilderness is admirable, or that it should be made useful, or that it is admirable to be useful and useful to be admirable. I have not mentioned wilderness aesthetics or wilderness sentimentality. I have not tried to persuade anyone that the howling wilderness is really a singing wilderness. Instead, I have argued that, howling or singing, it is a resource, perhaps the most important of all resources, both in its quality and its quantity, because it is essential for the long-range welfare of man. Nature is the final helmsman; no other reference point can be imagined.

THE GENETIC RESOURCES OF MAN

Curt Stern

THE GENE POOL

At present the quantitative aspects of human population changes overshadow all other population problems. The emergency of violent numerical growth of mankind requires measures independent of any considerations of the quality of the individuals which will constitute future generations. Yet, men are unequal in their genetic endowments and the corporate endowments of successive generations are bound to vary. These corporate endowments are designated as "gene pools," although genes exist not by themselves but within individuals.

As an approximation close to reality, it may be said that each individual carries the same total number of genes of many different kinds. Each kind is represented twice, one derived from each parent. It is not known how many different kinds of genes make up a human being, but there are speculative estimates. If two classes of genes are distinguished, those responsible for the synthesis of proteins including enzymes and those responsible for the regulation of the activity of the synthesizing genes, one may guess, on the basis of known facts about metabolism, that the number of synthesizing (or "structural") kinds of genes is of the order of 10,000. The number of the regulatory genes cannot be estimated at present. It may be much larger than that of the structural genes.

The gene pool of the structural genes of a world population of 3.5 billion is given by the product of 10,000 (kinds of structural genes) $\times 2$ (for the fact that each individual has two of each kind) $\times 3.5$ billions. This product is 70 trillion. This is a large but finite number. It is the given natural resource of the genetic endowment of mankind at one stage of its history. If the world population doubles by the end of the century, its pool of twice as many genes will have been derived from the present pool.

THE MOLECULAR NATURE OF GENES
AND THEIR ACTIONS

If each kind of gene were exactly the same in each individual, no problem of qualitative changes in the genetic endowment of successive generations would exist—barring mutation, the transformation of a gene into a different variant. However, each kind of gene may be represented by more than one type. Two types and frequently many may exist. It is possible to be relatively specific in this respect since the fundamental nature of structural genes is now well understood even if their individual constitutions are known only for some special cases. A gene is a chemical molecule of deoxyribose nucleic acid, DNA, whose structure resembles that of a ladder. The two uprights or backbones are made up of repeats of sugar and phosphate molecular groups. They are alike for all genes. The rungs of the ladder consist of two different kinds of pairs of molecular groups, the nucleotide pairs A-T and C-G, where these letters stand for well-understood, rather simple compounds. Each pair may be arranged in alternative ways, A-T and T-A, C-G and G-C. Thus there are four different types of rungs of the genic ladder. A gene possesses several hundred rungs of the pairs in a sequence which is characteristic for each gene. Thus, one of the 10,000 types of genes may be represented by

$$\begin{array}{c} A\ A\ C\ G\ T\ A\ T\ G\ T \\ |\ \ |\ \ |\ \ |\ \ |\ \ |\ \ |\ \ |\ \ | \\ T\ T\ G\ C\ A\ T\ A\ C\ A \end{array}$$

etc.; another type may be

$$\begin{array}{c} T\ G\ A\ C\ C\ C\ T\ T\ A \\ |\ \ |\ \ |\ \ |\ \ |\ \ |\ \ |\ \ |\ \ | \\ A\ C\ T\ G\ G\ G\ A\ A\ T \end{array}$$

etc. The number of possible combinations of the four double elements taken in groups of several hundreds is immense and permits each of the 10,000 types of genes, and many more, to have greatly different constitutions. The existence of varieties of each kind of gene is easily visualized by minor variations in its nucleotide sequences, such as a substitution of a T-A pair for one of its A-T pairs, or of a C-G pair for one of the T-A pairs, or of an inversion in a sequence of a few or many pairs, or of the duplication of a sequence or a loss of one or more pairs. Any given sequence has a very high stability during the persistence of an individual gene and in its replications, which lead to new genes that are exact copies of the preexisting one. Rarely however, an "error" leads to a change in a gene or in misreplication. The resultant mutant gene then has the same stability as the original one and replicates itself in its mutant form.

The specific sequence of nucleotide pairs in a gene is "translated" in the cells of an individual into a specific sequence of some twenty different kinds of amino acids linked together in a protein chain. Different types of structural genes thus account for the synthesis of very different proteins. Different variants of any one kind of gene may lead to protein chains which are essentially alike except for differences in a single amino acid, but the mechanism of cellular translation may be so strongly affected by a single nucleotide change in the gene that a highly abnormal protein chain may result. However, even if the protein chains controlled by two variants of a gene are very similar, the effect on the organism may be far-reaching. Thus the normal gene Hb^A for one of the protein chains making up normal human hemoglobin consists of 146 links. A variant of this gene Hb^S, which differs from normal by a change in a single nucleotide pair, controls a protein chain that is identical with the chain in normal hemoglobin of 145 links. The remaining 146th link, however, consists of a different amino acid. A person whose blood contains the variant hemoglobin instead of the normal one suffers from a type of anemia (sickle-cell anemia) which usually leads to a complex of illnesses and early death.

We may consider the gene pool of mankind or any of its subdivisions as the totality of all its gene kinds and variants, or we may separately consider the pool for any one kind of gene. If, for instance, this is done for the hemoglobin gene just discussed and if one regards the fact that more than two variants of the gene exist, one may speak of this specific pool as consisting of 3.5 billion pairs or 7 billion genes, most of which consist of the normal variant, the rest of the abnormal variant. The quality of the population is thus correlated with the proportion of the different variants in the pool of any one kind of genes.

GENES AND DIFFERENCES BETWEEN INDIVIDUALS

Differences among human beings are due to differences in genetic endowment, to differences in the physical, mental, and social environment in which the endowment expresses itself, and to interactions between heredity and environment. Tall people may be tall because they carry genes for tallness, or because they grew up under favorable nutritional influences, or because genes and environment favoring tallness coincided. These statements are axiomatic, but often it is not realized that every single human trait is subject to genic control. This is true of anatomical features which distinguish one normal individual from another, in respect to bones, muscles, nerves, and the size and proportions of whole organs. It is true of racially different features of skin, hair, and body build. It is valid for physiological properties: blood pressure, kidney function, metabolism of sugars, fats, proteins, blood coagulation, resistance to infections, longevity, type of ear wax, perception of colors, smell, orientation in space, and any other function. The two variants of a gene T control whether a normal person has the ability to taste a specific substance (PTC) as bitter or to regard it as tasteless. Another gene exists in variants, one of which leads a man to become severely anemic if he eats raw faba beans or is exposed to the pollen; given the presence of the variant, the same reaction may follow intake of the antimalarial drug primaquine. Ten per cent of American Negroes carry this variant, which is found also in other populations of non-African descent but is usually rarer in them. Still another gene occurs in variants, one of which permits use of the anesthetic suxomethonium and another which makes its use dangerous to survival.

There is a gene whose prevailing variety, P, controls the formation of a liver enzyme, which transforms the amino acid phenylalanine contained in every food protein into tyrosine. It has a variant, p, which does not lead to formation of the enzyme; so, instead of being converted, phenylalanine accumulates in the blood. (Since the excess of the amino acids is partially converted into a product which is excreted in the urine, the condition is called phenylketonuria.) The high concentration of phenylalanine in the blood causes damage to the brain of the developing child which usually results in severe mental deficiency.

Many genes influence mental performance less strikingly and yet partake in the causation of the wide range of mental attributes present in any population. The normal curve characteristic of numerous measurable traits of man has been interpreted in purely environmental terms. If there are ten alternative pairs of equally frequent environmental circumstances which lead to increase or decrease of a quantitative aspect of a trait, and if these circumstances are combined in random fashion without interaction, about one-thousandth of the individuals each will measure at the two extremes of the range, about one-hundredth each in the next, less extreme class and so on, with the majority forming the average class. Exactly the same distribution may be obtained when the environment is constant but when pairs of variants of five kinds of genes impinge on a trait. If the two variants of each pair are of equal frequency in the population and equally either add to or subtract from the end result and if the total effect depends on simple additive action of the genes and variants, then about one-thousandth of the individuals each will carry ten plus genes or ten minus genes, more will carry an unequal mixture of the genic agents, and the majority will be formed by those having five plus and five minus genes.

Such polygenic systems are well analyzed in plants and animals and their existence in man is certain, though a detailed specification is impossible at the present state of knowledge. Polygenic and non-genetic influences and their interactions are confounded in any actual case. This makes it difficult to partition their effects and excludes a precise assignment to that part which constitutes the raw natural resource of genetic endowment. Different geneticists may evaluate in very different ways the size of the genetic component of the variance of traits, but none denies the existence of such a component.

THE CAUSES OF GENETIC POLYMORPHISM

Genetically, the quality of a population depends on the proportion of the variants of each gene and their combinations. In large populations changes in these proportions can come about only by differential reproduction of carriers of different genotypes. Before discussing such future changes it is appropriate to try to answer the question why all populations contain more than a single variant of each gene.

Consider the gene P cited above which controls the presence of a normal liver enzyme and its variety p which causes mental deficiency. Specific individuals, all of whom have two representatives of the gene, are either PP, Pp, or pp. In the past whenever a pp child was born, its incapacity usually precluded later reproduction and the gene pool of P and p became changed by the loss of two p genes. In the course of centuries and millennia this should have depleted the pool completely. Yet the p gene is still there. A possible reason for this is mutation. Not only are the p variants selected against but they may also be created anew by mutation from P to p. If about one P gene out of 50,000 were not transmitted as such but as a mutated p, a population would be in a dynamic equilibrium; the loss of p variants owing to nonreproduction of pp persons would be compensated by their gain from mutation in PP or Pp persons.

There is observational evidence for the existence of selection-mutation equilibria accounting for genic polymorphism. As an exclusive explanation, however, this scheme has been challenged. It used to be assumed that the reproductive performance of mentally normal Pp persons either does not differ from that of PP persons or, possibly, is somewhat lower. Either way, the appropriate amount of mutation should account for the present coexistence of P and p in the gene pool. What, however, if Pp persons were slightly *more* fertile than PP's? In that case their higher reproductive rate would raise the pool frequency of p and might counterbalance the loss of p from pp individuals. It is difficult to test this hypothesis directly. Since there are many more Pp than pp persons in a population, it would take only a very slightly higher fertility of Pp than PP to compensate for the loss of p from pp. Such slight fertility differentials are hard to measure even in large populations. If they do occur they provide an alternative explanation for the presence of more than one variant of a gene in the population. Regardless of the occurrence of mutations, an equilibrium will be established between selection *against p* from pp and *for p* from Pp, thus keeping p in the gene pool.

The condition of a higher reproductive rate in persons with two different variants of a gene than with two like ones is called heterosis or overdominance. One of the best examples occurs in man. The gene variant Hb^s for sickle-cell anemia is highly deleterious to individuals who possess it in duplicate. Nevertheless, it occurs very frequently in certain African, Mediterranean, and Asian populations. If the high frequency depended on a selection-mutation equilibrium, the rate of mutation would have to be many times higher than that of all other known rates in man. However, the assumption of such a high rate of mutation is not necessary, since it has been shown that heterosis of carriers for both the Hb^s and the normal Hb^A variant of the gene provided the mechanism for maintenance of Hb^s in the population. Specifically, Hb^sHb^A persons are more resistant than Hb^AHb^A individuals to severe types of malaria. Thus, in regions of endemic malaria the better survival rate of the Hb^sHb^A group than for either the Hb^AHb^A or the Hb^sHb^s group led to a high proportion of Hb^s variants in the Hb gene pool. It is noteworthy that this advantage of the Hb^sHb^A group disappears when malaria is controlled. With this change in the environment, selection against the sickle-cell anemia Hb^sHb^s persons continues, not counteracted by heterosis and the Hb^s variant will decrease in frequency to a very low level, which may then be held by recurrent mutation pressure.

It is a matter of lively discussion at present which of the two mechanisms, mutation or heterosis, is more frequently involved in maintaining genetic polymorphisms. There is no general agreement. It may very well be that the example of the Hb^s gene is exceptional. One may speculate that gene variants which in double dose cause severe abnormal effects are unfavorable also in combination with a "normal" variant, and that in contrast mutant variants which in double dose cause only very slight effects may be heterotic in combination with a normal variant or with other variants that have slightly abnormal effects. The presence of two variants in an individual with more or less normal similar effects may give his biochemical machinery a greater versatility than the presence of either one alone.

The ultimate impact of changes in the frequencies of gene variants brought about by human activities clearly depends heavily on the relative importance of mutation versus heterosis. Heterosis provides a more powerful stabilizing tendency toward retaining a polymorphism of gene variants as compared with the weaker tendency of mutation to replenish abnormal genes.

THE IMPACT OF THERAPEUTICS

Medical progress has often resulted in improving the reproductive fitness of individuals who without treatment would have left no offspring or fewer than the average. In former times many hemophiliacs died before reaching maturity and thus did not transmit their deleterious gene. Now, blood transfusions permit longer survival, marriage, and reproduction. Severe cases of harelip and cleft palate often resulted in infant death or, if compatible with survival, to disfiguring appearance which reduced the chance of marrying. Now, successful surgery may permit survival and restore normal features. In diabetes, insulin has greatly improved the reproductive rate of affected individuals.

These medical measures are bound to change the frequencies of the inborn errors of metabolism, organ function, and development which are responsible for much illness. Granted that their appearance often depends on environmental factors as well as specific genes, the increased reproductive fitness of affected individuals will result in a change in the gene pool. Striking evidence for this statement is provided by an English study on pyloric stenosis, a condition in the newborn in which the opening from the stomach to the intestine is constricted by excessive muscular growth. It often proved fatal until, some forty years ago, an operation was devised to release the constriction and enable the patient to live a normal life. A study was made of the frequency of the defect among the children of persons who had undergone the operation when they were babies as compared with the frequency in the general population. In the latter, about three affected in 1,000 children were born; in the former 60 in 1,000.

Some formerly unfavorable genotypes are being transformed into neutral ones by therapeutics. Eyeglasses, for instance, compensate so satisfactorily for a variety of inherited imperfections of the optical apparatus of the eye that the unfavorable gene variants have lost their adjective. Similarly, insulin has largely nullified the detriment of diabetic constitutions. Cleft palate and pyloric stenosis, however, still impose serious problems on their carriers, since surgery is of a major nature. Therapeutic improvement of formerly low reproductive fitness of gene variants is bound to increase their proportion in the gene pool of the population. Therewith, therapeutic measures increase the load on society by having to provide, in each generation anew, remedial measures. If the abnormal gene variants were formerly kept in the population by mutation pressure, then their perpetuation by therapeutic means will increase their frequency in proportion to their being saved. If, on the contrary, heterotic advantage is involved, the relative increase in number of the abnormal variant will be less.

It is obvious that the impact of therapeutic measures on future generations depends on the persistence of a technologically advanced society which can provide eyeglasses, drugs, and medical services in increasing amounts. With such provisions society may well be able to support a larger and larger fraction of people with genetic imperfections. Nevertheless, theoretically a level may be reached at which the therapeutic load becomes too heavy—not to speak of the undesirability of undergoing excessive surgery and other treatment.

THE IMPACT OF GENETIC COUNSELING

If a pair of normal parents has been unfortunate in having produced a mentally deficient child of the *pp* type because of abnormal phenylalanine metabolism, they will ask whether the defect could recur in future children. A geneticist can tell them that the birth of the *pp* child indicates that they both carry the *p* variant in addition to the *P* variant which accounts for their being normal themselves. The probability of another *pp* child from two *Pp* parents is one in four. They will often decide not to take this risk. Until recently the discontinuation of reproduction by such parents would have had no consequences for the composition of the gene pool, since *pp* children would not have reproduced and thus would not contribute their *p* variant to the next generation. Most recently, a special diet, low in phenylalanine, makes possible a higher mental development; so *pp* children may be expected to reproduce and add their *p* genes to the pool.

A new problem arises when normal relatives of *pp* persons become concerned regarding their chance of having affected children. In many genetic diseases medical tests are providing means of distinguishing between normal carriers of an abnormal gene variant such as Pp and persons free from it, such as PP. Whenever a Pp constitution is suspected or known to exist a counselor may advise against marriage with another Pp person. This would result in the abolishment of selection against pp at the cost of an increased frequency of Pp in the population. If the frequency of p had in the past been held in equilibrium by mutational replacement, then the new type of assortative mating with its avoidance of $Pp \times Pp$ marriages would lead to a steady increase of p in the gene pool owing to the unchanged recurrence of mutations. If the former equilibrium was due to heterosis, an increase of p would occur likewise. In either case the frequency of Pp individuals would steadily rise until at some very distant time Pp persons would become so numerous in the population that the avoidance of marriages among them would no longer be feasible. Counseling in respect to the genotype of prospective marriage partners does not change the gene pool of a population, but for some sequence of generations it is potentially able to control the distribution of gene variants so that affected children do not originate. This, however, leads to changes in the equilibrium of gene frequencies and eventually to new problems.

THE IMPACT OF NEW MUTATIONS

Ever since the advent of nuclear weaponry, fear of the induction of gene mutations has been with us. Actually, the discovery in 1927 that X-rays cause mutations immediately led geneticists to discuss the danger to the resources of human genes. The danger arises from the expectation, confirmed by experience, that the overwhelming majority of new mutations is deleterious. This follows from the fact that the existing gene pool of any population is the product of natural selection which has accumulated a majority of gene variants with reasonably high fitness. Since gene mutations are more or less random changes in the genic molecule, the likelihood that they lead to increased fitness, or at least not decreased fitness, is small.

It is often argued that evolution depends on the accumulation of mutations and that this implies the existence of nondeterious mutations. In a way this is true, but there are at least two aspects which weaken the argument. For one, the price of evolution is extremely high. The new mutations which it accepts are a very small fraction of those it rejects. If some radiation-induced mutations indeed confer on their bearers a higher fitness, a hundredfold larger number will have the opposite effect. A second aspect concerns the type of mutations which enter promising evolutionary pathways. Most such mutations have very small effects by themselves, since the probability of their being compatible with normal fitness depends on their property of not upsetting the usual course of development. Evolution by means of mutations as raw material is thus not a progression by striking jumps but by slight changes in some genes which offer the opportunity of exploring new fitnesses by trial of the new variants in manifold combinations with the variants of other genes already present in the gene pool.

Deterioration of the human genetic material by irradiation is not to be feared alone from nuclear weapons and the fallout which results from nuclear testing. The genes of many persons have been exposed to irradiation as a consequence of diagnostic and therapeutic treatment by X-rays and other ionizing radiation. In recent years the amount of exposure has been greatly reduced without interfering with legitimate medical purposes. Lower radiation dosages and better shielding of the ovaries and testes has undoubtedly diminished the number of induced mutated genes transmitted to future generations.

Radiation is not the only cause of mutations. Mutations have been shown to occur "spontaneously" whenever suitable tests have been made in whatever microbe, animal, plant, or man. Few of these can be due to natural irradiation of genes from radioactive atoms in the environment and in food, or from cosmic rays. Most of them must be caused by accidents which change the chemical constitution of a gene or result in an error during its replication. Such accidents are presumably of chemical origin. We now know of many substances which are mutagenic, and their interactions with the gene molecules begin to be understood. Many of the most effective mutagens used experimentally do not occur naturally in an organism, but the probability is high that natural products of cellular metabolism may occasionally effect a chemical change in a gene. It may be assumed that some of the many natural and artificial substances which enter our body as food, drugs, stimulants, and irritants are mutagenic. Even if individually their mutagenic actions are rare, the sum of their effects may be far greater than that of radiation. This leaves out of consideration the mutagenic effects of a nuclear holocaust.

The quantitative effects on the population of the induction of mutations by man-made radiation and by chemical mutagens depend on the relative importance of the two main methods by which genic polymorphism for genes unfavorable in double dose is maintained: mutation and heterosis. As suggested above, this relative importance may be different for gene variants with slightly subnormal effects in double dose and for those with serious effects. An increased mutation rate will add to variants both with minor and with major effects. Many of the minor variants may already be maintained by heterosis; the major ones will simply add to the load of severe mutations.

THE IMPACT OF REPRODUCTIVE DIFFERENTIALS WITHIN AND BETWEEN POPULATIONS

For many decades of the present century as well as earlier the various socioeconomic groups in Western societies differed in their mean reproductive rates. Specifically there was an inverse relation between the reproductive level of a group and its position in the social scale. The professional and managerial levels produced significantly lower numbers of children per adult than the less skilled and unskilled levels. It was established that the mean performance on intelligence tests of the children from the first-named groups were higher than that of children from the second-named groups. It was further established to the satisfaction of many though not all students that the differential test performances could not be accounted for solely on the basis of social advantages and disadvantages. Granting the validity of these findings, it was concluded that the differential reproductive performances at different socioeconomic levels implied a loss of gene variants for higher intellectual potentialities.

The consequences of such "gene erosion" within populations would be a progressive lowering from generation to generation of the mean intelligence test performance. Various estimates of the decrease in the mean score of a population were made. They predicted a drop of the mean score of between one and five points per generation.

The uncertainties of such estimates are very great, for they depend on assumptions about the degree of genetic differentials between socioeconomic groups which are at present nonprovable. It seems reasonable, however, to assume that in societies with social mobility some differentials exist. Social mobility tends to stratify people in different layers according to traits such as mental genetic endowment. Even though serious delays, extending over generations, keep relatively incompetent persons of upper layers from falling and highly competent persons of lower layers from rising, upward and downward movements characterize most societies which do not have rigid caste systems.

The hypothesis of declining mean mental endowment owing to differential reproductive rates has been tested by studies (Scottish Council, 1949) on two groups of Scottish school children, all those (more than 87,000) who were eleven years old in 1932 and all those (more than 70,000) who were of that age in 1947. The results were surprising. Far from showing a decrease in the mean score of a verbal intelligence test there was actually a significant rise, from 34.5 in 1932 to 36.7 in 1947. Higher scores were obtained for both boys and girls, but the increase was far greater for girls.

The comparison between the score improvement of the two sexes suggests strongly that nongenetic differences between the two populations of eleven-year-olds were at work, since genetic theory would predict very similar endowments of the two sexes in the same population. Now, if nongenetic factors were active in influencing the scores between the sexes they may have been of importance also for score differences between the 1932 and 1947 populations. The nature of these external factors remains unknown. "Test sophistication" acquired from greater familiarity with mental testing, or earlier maturation, may have affected the scores, as well as more subtle social differentials of motivation, family structure, and accessibility to general information.

Was there then no decline in genetic endowment in spite of differential fertility? The data do not provide an answer. There is no obvious reason why the observed rise in performance should have had a genetic component. If, then, nongenetic factors led to an improvement in scored performance did they produce this effect on a genetically unchanged basis or was the observed improvement accomplished in spite of an actual decrease in gene resources?

The expectation for a trend toward decreasing intelligence of societies with inverse reproductive differentials between socioeconomic layers is based on the assumption that no differentials are present within the layers or, if present, that they reinforce between-layers differentials. There is some evidence from American and European studies that at least within some groups reproductive rates and higher performances are positively correlated. Some recent studies, limited in extent but nonetheless highly noteworthy, have been concerned with the fertility of individuals classified by intelligence test performance independently of their socioeconomic status. Data on the completed fertility of women whose IQ scores were available from the time they had been at school have indicated that not only the fertility of the poorest test performers is high but also that of the best performers. In other words, the relation is not inverse throughout, but shows a bimodal distribution. Calculations indicate that in these sample populations the mean IQ range holds its own reproductively. Furthermore, it is possible to construct genetic models in which a lower than average reproductive rate of highly endowed groups is counterbalanced by a lower than average rate of abnormally poorly endowed groups such as imbeciles, with the large medium group having the highest rate. Such models fit a state of affairs in which no gene erosion takes place.

In several Western countries and perhaps in other parts of the earth also, the differentials in reproductivity of different population layers have greatly decreased in recent decades. This is due primarily to the spread of family planning from selected to all socioeconomic groups. Whether the differential will disappear completely is hard to foresee. It seems, however, that our concern for the maintenance of our mental genic resources need not be so urgent as we had feared. Even had there been continuing processes toward loss of genes involved in intelligence scoring, the polygenic basis of this quantitative trait assures that a vast reservoir of "high" genes is present to a diluted degree in individuals with lower performance, a reservoir from which for a long time each generation would reconstitute concentrated high endowments.

THE IMPACT OF REPRODUCTIVE DIFFERENTIALS BETWEEN RACES

It has been estimated that in 1650 the number of African and Asian persons was four times that of Europeans and their derivatives in other continents. Three hundred years later, the ratio had decreased to about 1.6 to 1 and by 1980 it is likely to have risen again, to be 2 to 1. Changes of this type constitute equivalent changes in the gene pool for genes that differentiate anthropologic groups, such as genes for certain anatomical features and pigmentation. They also imply changes in the pool for gene variants that have high but different frequencies in different groups. Thus, variants for blood-group substances O, A, and B exist in most human populations, but the variant for B makes up less than 10 per cent of the pool in many European populations but is closer to 20 per cent in Asian and African groups.

Some of the existing differences between human races have probably come about by initial chance processes. Others are the resultants of natural selection which adapted groups of people to specific environments. Presumably many specific racial traits have lost their biological significance, and changes in the frequencies of the gene variants controlling them are without importance. The question remains whether there are genic differences between races to which a valid value judgment can be applied. Indeed, it is very likely that gene variants exist in racially variable proportions for such traits as resistance to infectious diseases and susceptibility to specific types of tumors, as well as for physiological properties related to climate and other aspects of the physical environment. Given specific environments, specific gene variants constitute valuable resources.

A broader question relates to the possibility of differences in mean mental endowments between racial groups. This question is so loaded with prejudices past and present that even to raise it is often regarded as improper. Nevertheless, many geneticists believe that such differences in averages exist, since it is very unlikely, a priori, that the polygenic systems which underlie normal variation in most traits, in each population lead to exactly the same mean performance. These geneticists profess that it remains unknown for any two racial groups which of the two has an endowment for higher performance. There are no available tests which can be applied with equal validity to two populations with different social and cultural milieus. Different cultural achievements are not measures of innate endowments and, in any case, are not easily subject to ranking.

Whatever genetic differences exist, there is no doubt of the very wide overlap in the distribution of endowment. Many persons in the more highly endowed group will be inferior to many in the less highly endowed group. Yet, even with slight mean differences, there would be considerable disparity between the two populations in the frequencies of persons with very high or very low scores. Whether mankind needs more than a small number of very highly endowed individuals, and whether this number is already reached within a population having the lower average endowment, is at present not a matter for objective judgment.

Social and cultural factors of nongenetic type are of overwhelming importance at the present stage of mankind. Therefore, discussion of genetic factors is a subsidiary undertaking. It would seem, however, that suppression of such discussion would tend to distort one's view. Gradually the equalization of opportunities should enable all men to realize their potentials. Concomitantly, the existence of genetic diversity within and between human populations will not only become more obvious than at present but also demand intense consideration.

THE POTENTIAL IMPACT OF EUGENIC MEASURES

Proposed measures for improvement of the genic resources of mankind have been both positive and negative. The former have as their goal the increase in gene variants which may determine traits better than the average. The latter hope to reduce the frequencies of gene variants which have clearly deleterious effects. Negative eugenics may be characterized as preventive and positive eugenics as progressive.

It is not only easier to define abnormal traits compared with better than normal ones. It also is simpler to discuss preventive than progressive eugenics. Many strikingly abnormal genetic traits are caused by single abnormal gene variants, whereas most traits of supernormal nature are controlled by multiple combinations of many different gene variants each of which by itself has only a slight effect. Selective measures against single strikingly abnormal variants require the imposition, voluntary or involuntary, of restrictions on reproduction of a small minority of individuals, but such measures in favor of individuals at the upper end of a normal distribution curve involve interference with full reproduction of a majority of the population.

The speed with which a population can be cleansed of gene variants that have strikingly abnormal effects is different for "dominant" variants which cause their severe deleterious effect when present in single dose, and for "recessive" variants which must be present in double dose. The dominants can be recognized in the individuals which possess them; the recessives are hidden whenever they occur in single doses. In neither case is the "cleansing" process as effective as a first guess would suggest. Individuals with a dominant gene for severe defect usually are strongly reduced in their reproductive rate. Hence the selection-mutation equilibrium implies that most affected persons are not the offspring of affected parents but the result of new mutations. Thus, after reducing or excluding reproduction of the defectives in each generation, there will still be born those loaded with a newly mutated abnormal gene. While a single generation of selection against a dominant gene may reduce its presence appreciably, no further progress can be made.

The opposite situation holds for recessive genes with severe effects in double dose. Since, as can be shown empirically and deductively, the majority of such genes are hidden in individuals who have both a normal and an abnormal gene, selection against the afflicted individuals with double dose controls only a minority of recessive variants. Nothing can be accomplished by selection against persons affected by such conditions as phenylketonuria (pp), which do not permit reproduction in any case, and are reconstituted in each generation by marriages of normal persons each of whom carry one recessive gene ($Pp \times Pp$). Some progress by preventive measures can be made against recessives which biologically permit reproduction, but the decrease in numbers of the afflicted is slight after one generation of selection and becomes less and less effective in successive generations. Complete exclusion from reproduction of all recessively affected persons of a condition occurring in ten per thousand of a population would not prevent the appearance of more than eight per thousand in the next generation, and to reduce it to one per thousand would require twenty-two successive generations of such selection. To make a further ten-fold progress in the reduction of affected, that is, to have only one occur among ten thousand, would demand sixty-eight additional generations of selection. Much more can be accomplished by cessation of reproduction of parents to whom a recessively affected child has been born, and such action would indeed help to cleanse our genic resources of defective variants.

Selection for traits above the average of a population has been successful in animal and plant breeding. Given specific goals such as faster growth in pigs, greater speed in horses, higher yield of milk in cows, of eggs in chickens, or of seeds in corn, breeders have artificially created strains which outproduce by far the original wild forms which had been provided by natural selection. Theoretically, similar results could be obtained in man. To quote a recent statement by a distinguished geneticist: "Our experience with other species suggests that an appropriate scheme of intensive selective breeding would take no more than a dozen or so generations to double or halve either the average stature or the average level of IQ, according to the direction in which we applied the selection." It would not be possible, however, to select simultaneously for a great number of desired improvements, and separate programs of selection would result in separate strains. Even more strictly than in present societies with rigid caste systems, such strains would have to be continued as an assortment of numerous noninterbreeding populations living side by side.

The genetic feasibility of such schemes is not equivalent to their social feasibility, and the ethical problems they raise are exceedingly far-reaching. No one aware of the inherent dangers would contemplate any but voluntary attempts at progressive eugenics, and many would shrink even from such restricted action. Others, like H. J. Muller, are advocates of partial measures to improve the genic resources of mankind. They propose the use, in artificial insemination, of sperm from donors selected on the basis of special achievements. Even more radically new methods have been envisaged, such as the cultivation of egg- and sperm-producing tissues in test tubes and the implantation of fertilized eggs from such cultures in the uteri of women desirous of nurturing a potentially exceptional child. It begins to be possible to fulfill the dreams of a brave new world.

There are other means in sight of improving the genetic resources of man both by decreasing deleterious variants and by upgrading quantitatively varying traits. Direct changes of defective human genes into nondefective ones, while not yet accomplished, seem not an impossibility. Such "gene surgery" would abolish much suffering and make unnecessary even the thought of harsh selective programs. In the meantime biological and social measures can be expected to improve the working of our present genic resources. Three or four generations are equivalent to a century—a short time in which to make possible striking changes in the gene pool, whether negative or positive, but a long time for the discovery and application of methods to make the best out of the genes we have.

REFERENCES

Mather, Kenneth.
   1965. *Human Diversity*. New York: The Free Press. 121 pp.
Scottish Council for Research in Education.
   1949. *The Trend of Scottish Intelligence*. London: University of London Press. 151 pp.
Sonneborn, T. M., ed.
   1965. *The Control of Human Heredity and Evolution*. New York: The Macmillan Co. 127 pp.
Stern, Curt.
   1963. *Principles of Human Genetics*. 2d ed. San Francisco: W. H. Freeman and Co. 732 pp.

Part II   QUALITY ISSUES IN THE
MANAGEMENT OF SOME MAJOR
RESOURCE CATEGORIES

,

PROBLEMS OF QUALITY IN THE
PRODUCTIVITY OF AGRICULTURAL LAND
William A. Albrecht

In the discussion of our natural resources and processes, guidelines are essential. The usual quantitative characterizations of land contribute little to suggest the quality of agricultural products that may be grown on it. To say that the earth has two and a half billion acres of tillable land means little until equated against the earth's human population of some three billion inhabitants. This tells us that each of us has, in the mean, something less than an acre of arable soil for growing food. The productive quality of one acre of the earth's surface, in terms of its yield of protein in life forms and of foods required by them, becomes far more significant than its quantity as mere surface area. The significance of soil quality is made clear when we realize that an acre does well to grow the beef equivalent to the present meat consumption per person per annum in the United States.

The term "meat" may well represent all living and self-multiplying substances characterized by their protein content. Proteins are provided for us in high concentrations in animal tissue, or in animal products which have food value for other forms of life: milk, cheese, eggs, and so on. Our thinking progresses from the quantitative to the qualitative, from the amount of land in acres to the productive capacity of the soil, and then to what it grows in terms of quality in nutrition for *healthy survival,* which becomes the criterion of quality in production.

Dimensional descriptions of the soil tell us little of its biological importance. They fail to bring this unique natural resource into focus as a dynamic natural substance that takes its origin from the lifeless inorganic rock minerals at the earth's surface. Our quantitative evaluations of soil have, to date, not brought about sufficient appreciation of the soil to assure its conservation. But when we remind ourselves that the soil determines the biosynthetic processes of all that is organic and lives on a given area, its third dimension—the depth of fertile surface horizons in the profiles of rock fragments and organic matter, weathering and decaying under climatic forces—becomes highly important. A fourth and a fifth dimension are important also: time, in eras and centuries; and energy, and its transfer finally into heat to escape into space, in accordance with the second law of thermodynamics. The combination of all five dimensions emphasizes the qualitative and dynamic aspects of the soil in the support of life, including the exhaustion of its fertility, its possible renewability, and its potential support of the different trophic levels, or life forms, in the entire biotic pyramid built up by, and dependent on, the soil.

As another emphasis on quality in production for the support of life we shall consider the soil, not so much for its role in giving us carbohydrates and fats as energy foods, but rather for its support of the biosynthesis of proteins. The quality of proteins needs also to be considered according to their quantities, not of nitrogen measured by ignition, but of their constituent amino acids, each in the amount required for health by any living species.

For additional refinement of the concept of quality in the productivity of the soil, we may well consider the evolution and survival of each of the several species in their ecological array. That array may be interpreted as nature's report on the natural biological assay for success by a given species in each soil area. By recognizing the special nutritional requisites of species, correlated chemically and biochemically with the soil properties in its geoclimatic setting as potential production of proteins, we can do much through our treatment of the soil to raise the quality of production. We can at least prevent ecological misfits or violations of the bioclimatic laws and bring about fuller accordance with them.

SURFACE PHENOMENA GIVE SOIL ITS DYNAMIC QUALITIES FOR PLANT NUTRITION

The simple fact that the soil is the surface layer of the earth makes it a unique resource. The surface aspect of every kind of matter became a part of basic science with the advent of colloidal chemistry, which deals mainly with "surface phenomena." These represent the forces of concentration of any two kinds of matter at their interface or area of contact. In that concept, the surface of the earth, as a bit of cosmic dust in contact with the atmosphere, sets up the many forces concentrating into maximum densities the several components of the atmosphere, and likewise those of the soil, at the soil-air interface. The plants inserted there take advantage of all these for production of their vegetative mass and for survival.

The concept of the variation in climatic soil development as the major determinant of biotic geography is now coming to be accepted. The climatic pattern of soil development of the United States illustrates its operation. The increase in rainfall from the Great Basin eastward as far as the midcontinent brings increasing kinds and quality of vegetation as water weathers the rock into deeper soil with more clay and more fertility for plant nutrition through protein production. From the midcontinent eastward there is a reciprocal decline in protein production, although the clay content and its colloidal behavior increase. The monovalent cation, hydrogen, increases but the divalent cation, calcium, adsorbed on it, decreases.

With that decrease in adsorbed calcium of the soil, the percentage of calcium and of other cations in the dry matter of the vegetation decreases. But the relative decrease in calcium is much greater than that of other elements. Consequently there is a narrowing calcium-potassium ratio. The decline in protein production connected with the calcium is more rapid than in the carbohydrate production with which the potassium is related in the plant's processes. Thus a shift from a semi-arid to a humid condition causes fertility depletion in the soil and decline in nutritional quality, especially in proteins and the inorganic essentials associated with them in the vegetation.

The same pattern of changing quality of vegetation according to the degree of soil development from arid to moderate rainfalls and from moderate to excessive ones is seen in particular traverses in the Soviet Union; in Australia, in the changing qualities of wool; in Argentina, in grasses for growing and fattening grazing animals; in Hawaii, according to altitude rather than longitude and latitude; and in Java, according to recentness of volcanic eruptions.

Calcium, and its distribution in the profile, was the first index by which the soil surveyor correlated soil quality with rainfall and temperature. The student of crops mapped the virgin vegetation as grasses in the western parts of the midcontinent and as forests in the eastern parts. He spoke of prairie soils and forest soils. That led to the erroneous implication that the soil is the result of the vegetation rather than the cause of it. The maps of animal production as patterns of protein potential of the soil show that high-protein animals, like cattle, are grown in the western midcontinent. But pigs, as pork, for speedy gain in weight through fattening procedures (abetted by castration) are grown in the eastern midcontinent. The highest percentages of land under cultivation are found in the midcontinent, where animal population densities are also at their maximum.

In the correlation between soil fertility for protein production and biotic geography, the former is the major cause, the latter the effect through soil chemistry and biochemistry. In maps of electrical transmission for radio reception, the "excellent" and "good" areas in the midcontinent almost mirror the map of soil fertility. Poor reception by radio correlates with minimal qualities of health and survival outside the midcontinent area.[1]

When the atmosphere serves as one line of electrical transmission by wave for radio, and the soil is the other conducting line of the circuit, we need to be reminded that the soil will be a good conductor only when both moisture and electrically active ions of the fertility elements are ample. Neither water without the ions nor the latter as salts in a dry soil are effective conductors for radio reception. Nor will they provide the chemodynamic services in the soil's production of quality feeds and foods. Ions serve in the soil, not as high concentrations in the salts of arid and alkaline regions, but as elements of low degrees of ionization adsorbed on the soil colloids and remaining there in the face of higher rainfalls that leach salts out of the soil. We are slowly organizing the biochemical and ecological facts that prove the soil to be a prominent determiner of the quality of crop growth.

The better understanding of surface phenomena active within the soil has clarified our understanding of how soils can resist loss of nutrient elements to percolating waters, retaining them in ample supplies readily available to plants. The soil's finer particles of clay represents enormous total surface in very small mass. Surface phenomena serve to concentrate, on and within the clay, some of the nutrient elements weathered out of the rocks. These are held by the clay against loss to percolating water. But those nutrient elements, although chemically insoluble, are nevertheless biologically available to plant roots. Through the weathering process of the reserve rock, inert, inorganic matter becomes creative according to its total exposed surface rather than to its quantity as weight. A shift in emphasis from *static mass* to *dynamic surface* shortens time and magnifies energy, thus emphasizing quality over quantity as a criterion of soil classification.

Grinding rock into particles no smaller than silt size (0.02–0.002 mm), carried and deposited by the winds, gives sufficient increase of surface and of rate of weathering in contact with acid clay to improve

---

[1] William A. Albrecht, "Soil Fertility and Biotic Geography," *The Geographical Review*, Vol. 67 (1957), pp. 86–105.

crop growth.[2] The virgin silt soil (loess) of the midcontinent serves as a regularly renewed resource with its half-ton deposits per acre per year of unweathered minerals picked up from the winter-dry Missouri River bottoms and plains for deposition under higher rainfalls to the northeast. In this way the protein-producing potential of crops has been not only maintained but extended in area.

Plot tests on different sizes of limestone particles have demonstrated the extended nutritional service for the second year's growth of sweet clover when the particles were nearer ten-mesh than hundred-mesh size. Larger particles, broken down by reacting with the clay, gave fewer foci offering calcium as plant nutrients, yet gave saturation to higher percentages of the soil's adsorption capacity and thereby larger amounts for the nourishment of the roots. The smaller particles, providing more numerous foci in the soil and thus promoting the speedy adsorption of all the calcium, left the clay holding that element at high energies beyond the competitive power of the roots to remove it for plant nutrition in the second season. Gradual nutrient release by slow reaction rather than speedy delivery as in soluble form is a basic natural principle in the growing of high-quality agricultural products. It emphasizes nature's use of rock fertilizers rather than soluble, highly concentrated salts for maintaining the soil as a resource for growing microbes and plants.

These facts support the concept that the soil need not be a uniform medium, but may well be a heterogeneous collection of foci of minerals or rocks weathering while in contact with clay and plant roots. Plant growth, then, is a summation of all these centers of fertility as the roots move to and get from them all that is needed for maximum quantity and quality, or fail accordingly. By this concept, both the more soluble and the less soluble nutrient materials applied in granular or fragmental forms will better maintain this seemingly beneficial heterogeneity of fertility sources than would any practices aimed at reducing the soil to the uniformity of nutrient solutions.[3]

Clay serves as the negative anion in the surface phenomena of adsorbing the positively charged elements (cations) such as hydrogen, calcium, magnesium, potassium, sodium, and the trace elements. Extensive tests of the retention of cations by soil colloids (clay and humus) have shown that the varying ratios of cations, in the total ex-

---

[2] E. R. Graham, "Soil Development and Plant Nutrition: Nutrient Delivery to Plants by the Sand and Silt Separates," *Soil Science Society of America, Proceedings*, Vol. 6 (1941), pp. 259–262.

[3] William A. Albrecht, "Plant Nutrition and the Hydrogen Ion: Relative Effectiveness of Coarsely Ground and Finely Pulverized Limestone," *Soil Science*, Vol. 61 (1946), pp. 265–271.

change capacity of the soil, are reflected in the varying chemical compositions and qualities of crops.

Of the cations or nutritive elements listed above, only hydrogen is not taken by the plant from the soil. Vegetation varies greatly both in total protein content and in its amino acid constituents according to the ratios or percentages of saturation of the colloids by the exchangeable cations. The ratios depend on the differing degrees of soil development as well as on many other factors. They provide a refined means for interpreting the diet offered to the plant by the soil. But there remains much of plant quality which surface phenomena will not interpret, however helpful that concept has been.

The chemical composition of soil may represent either a few or many inorganic elements of rock and mineral origin. Analytical inventory of soil elements by ignition is but a gross measure of the adequacy of inorganic nutrition for microbes, plants, and higher forms of life.[4]

INTERDEPENDENCE OF SPECIES EMPHASIZES EQUILIBRIA FOR THEIR NATURAL CONSERVATION

There are interrelations representing conservation between the several interdependent trophic levels. The remnants of any resource, the organic and inorganic wastes from one life form, may be the major resource of another in terms of matter and energy. That relation may be as intimate as symbiosis. Soil resources, comprised of varying geological matter weathered under differing climatic forces, must of necessity vary widely according to geographic location. Within limited areas the soil attains brief states of equilibrium which is characterized by a particular productivity. Over eras of time, the soil is never in

---

[4] Editor's Note: In discussion, G. B. Bodman, Emeritus Professor of Soil Physics at Berkeley, emphasized the fact that the quantity of crop produced is in no small measure correlated with the physical properties of the soil. Deeper soils are correlated with larger yields, but, owing to their gentle relief, are only too frequently withdrawn from agricultural use and become sites for airfields and factory, housing, and highway development. He pointed out that the yield of dry matter of several important field crops has been shown to be affected by the nature of the energy of release: soil water-content curves which serve as indicators of time of irrigation for a given soil. The postponement of irrigation until the lowest limit of water availability to the plant has been reached may seriously affect yield. The "water-release curve" of a given soil is a soil quality, therefore, that is related to crop yield in irrigation agriculture.

Trace elements in soil are quality attributes. Several trace elements are known to affect the physiology of the plants grown on them and also the nutritive value of the plants to, and hence the health of, animals that consume the fodder. An instance was cited of molybdenum toxicity in the San Joaquin Valley of California which was traced to the form of the molybdenum present in the soil.

more than temporary equilibrium while on its way to solution and to the sea. The potential productivity of the soil must then be variable with time. Its productive quality only appears to be a constant, as when we speak of it simply as "land." Soils are the natural dynamic result of spending the innate and acquired energies of the earth which, like a wound-up clock, is in the course of running down.

To support microbial life, the lowest trophic level, the soil must provide the essential mineral elements and the organic substances which microbes oxidize for chemical energy, captured previously from the sun and stored by plants. Microbes are not equipped with chlorophyl, as are plants, to capture the sun's energy for themselves. The microbe is a single-cell form that may be used to illustrate the principles applicable to multicelled bodies. Visualize any sterile soil inoculated with life by a single microbe. Assuming all the essentials available in the soil for the microbe, the numbers of the increasing population, plotted as ordinate against time as abscissa, will give the characteristic sigmoid curve. Its shape suggests an S with the top pulled to the right while the lower left end remains fixed. During the early stage, the population increase is geometric. It doubles per unit of time of cell division. But soon the shortage in the food supply and the accumulation of wastes operate against continued increase. These two factors give the second half of the curve, which is an inverted duplicate of the reversed first half.

Thus a simple laboratory illustration can demonstrate how any life form comes into a favorable setting, runs its course to climax or maximum, and then passes on to extinction. This occurs unless it is able to modify, advantageously, its own original physiological processes. But while microbes change, the nature of the soil is also changed to the extent that it can no longer serve as the growth medium for another similar population. It could, however, be such for other microbes with different but *lesser* requirements. But as producers these would deliver *less quality* in their product. That principle applies to the soil in relation to any and all of the different strata of life when each goes through the stages of introduction or establishment, increase or growth to a climax crop, and disappearance through the same factors, forces, and phenomena cited for the microbe. We are prone to see only the second period, the increase toward a climax crop.

The struggle by multicelled bodies duplicates, in principle, that by the microbe, since the former are the summation of many cells combined in a larger body. The soil microbe carries out oxidation and reduction, thus gaining or spending combined and stored energy, much of which escapes as heat. The resulting acids—carbonic, nitric, sulfuric, and many organic acids—become weathering agents of rock minerals, thus increasing the availability of their nutrient elements. Microbes, in a sense, carry out analyses and syntheses. Their growth becomes a highly differentiated cell division only when certain compounds trigger that self-multiplying phenomenon, which transcends the mere increase in mass by living tissue. The microbe illustrates the same principles which operate in all higher forms of life, fitting themselves into the environment and surviving by growth and multiplication in accordance with the fertility of the soil.

Because of their dependence on organic matter for energy, through decay or oxidation, microbes constitute the decomposer stratum in the biotic pyramid. They obtain their requirements for growth or tissue building from both the decaying organic debris and the decomposing minerals of the soils. Many of the essential elements and organic nutrient compounds remain in cycles of use and re-use, since sulfur, nitrogen, phosphorus, and carbon are major constituents in microbial left-overs.

Microbes may reproduce at the rate of one generation per hour or less. They synthesize extra- and intracellular compounds which protect them against competitors and predators. Those chemical compounds may serve as antibiotics for the human body. They may be considered an evolutionary (accidental) good fortune, a by-product of value from metabolic wastes of the microbe. From such, in response to hypodermic inoculations, the human body generates antibodies.

The production of antibiotics by microbes suggests that the biosynthesis of organic compounds by each life form may provide support to others. It certainly emphasizes their interrelations. This is suggested for plants also when gardeners speak of the compatibility and protection provided by certain crops for others when they are interplanted. As decomposers and synthesizers of myriads of organic compounds, the microbes make for themselves an environment in what is often said to be a "living soil." Thus nature practices conservation by using the wastes from a lower form of life to make the environment more compatible with its survival and, incidentally, that of other and higher species as well.

Man as a species hardly manages his own environment in such a way as to actively encourage the survival of other forms of life along with his own. Instead, he more often destroys the equilibria on which the other species depend for survival.

Human societies, however, are giving more thought and effort to the conversion of city wastes for re-use as fertilizer. Such measures may also serve to prevent pollution by the many complex organic compounds and dangerous inorganic compounds that are dumped directly by industry into streams. The flushing of organic substances from cities into streams and the sea has been needlessly depleting our soils. As a result, agricultural production has shifted to one of increased bulk and reduced nutrition. We easily forget that the return of organic matter to the soil is crucial in the maintenance of the quality of products now flushed into the sea. It must become a major effort if the decline in quality of production is to be checked.

Plants, as producers of stored energy, are a major part of the biotic foundation. They are the only significant producers, relative to quantities of stored energy, within the biotic arrangement. Like the microbes, they are unique in that their metabolic waste of carbon dioxide in water gives carbonic acid with ionic hydrogen, a nonnutrient cation within the soil that is adsorbed on the plant root. There it exchanges for the nutrient cations—calcium, magnesium potassium, ammonium, and others—adsorbed on the colloidal clay and humus, and mobilizes them. Plants, as excretors of other acids and compounds, or as oxidizers and reducers, speed the breakdown of rocks and minerals into finer fragments to produce colloidal clay and thus make available the elements of fertility. Plants, much like the microbes, improve the environment for their own survival through addition to the soil of their wastes and by-products.

Plants are the producers for all life forms because they are nature's means of storing and distributing, as foods, in usable chemical forms, the energy supplies imparted to the earth by the sun. Such energy is collected by the unique process of photosynthesis. This is brought about by means of the inorganic part of chlorophyl to help that enzyme combine carbon, hydrogen, and oxygen into the common energy food, the carbohydrates. These are the plant's food for its own metabolism. In that respect, plants are the energy source not only for themselves but also for biochemical processes at various trophic levels.

Compounds of six carbons and their multiples or units near the figure six constitute most of the carbohydrates and the major part in the chemical composition of a plant's dry matter. While carbohydrates represent high energy values, reduction or removal of oxygen from those compounds pushes their energy values even higher in the straight-chain hydrocarbons, suggesting fats, or in those of benzene, a six-carbon ring structure. While the human body metabolizes the former compounds of limited length, apparently it is unable to break the benzene ring, which the liver, kidneys, skin, and excretory system as a whole must accept as overload. Nor is the benzene ring broken readily by microbial metabolism. Hence it persists from wood and bark through coal and oil to be recovered by their destructive distillation.

Carbohydrates are also starter compounds for the plant (and microbe) into which its synthesis of nitrogen, sulfur, and phosphorus yields the amino acids constituting proteins as living tissues which grow, protect, and reproduce. Thereby plants, nourished by the soil and its microbial aids, are in one sense the only producers. They are the sources of both food energy and growth substances synthesized from chemical elements and passed on, as life support, for all other populations, either higher or lower in the biotic pyramid.

In that schematic arrangement, the three lower strata of plants, microbes, and soil are the basic synthetic support of all living matter. The plants supply stored energy and growth substances. The microbes remove accumulated organic matter, serve as salvage crews to return water and carbon dioxide to the atmosphere, and recycle the elements of mineral origin and many organic substances for nutrition of plants and themselves within the soil. Plants live in both soil and atmosphere. Microbes live very much within the soil. They were the first forms of life to appear and will doubtless be the last to disappear. They are, however, often in symbiotic relationship with plants.[5]

Above these three, all life forms are, in the final analysis, dependent on the organic substances produced by plants and transformed mainly by simplifications at higher trophic levels. Were we able to catalogue all the major biochemical reactions in each higher life form, that knowledge could serve to permit each species, except man, to be its own bioassaying agency for us. Species other than man do not control the soil or environment. For these other species, from insects up through carnivores, the criterion of protein produced, including the required amino acids, aids in the assessment of soil fertility. Such assessments require increasing refinement as the biochemistry becomes more complex at higher levels of life. We need also to increase our knowledge of the biochemistry of plant nutrition, to extend our concepts of the soil's functions to encompass more of the non-

[5] Studies of the amounts of nitrogen and carbon in soils, especially their ratios, have established these as quantitative expressions of climatic potential for quality in agricultural production. See M. F. Miller, "Studies in Soil Nitrogen and Organic Matter Maintenance," *Missouri Agricultural Experiment Station,* Bulletin 409 (1947); Hans Jenny, "A Study on the Influence of Climate upon the Nitrogen and Organic Matter Contents of the Soil," *ibid.,* Bulletin 152 (1930). This is the first of several papers by Jenny on this subject.

ionic reactions, including the chelated compounds containing both inorganic and organic ash percentages approaching those of living substances. And we need to accept a *molecular biology*. Modern man does not yet submit to the use of his body's biochemistry as a bioassay agency for measuring the soil's productive capacity. Instead, he aims to modify the environment to his desires, many of which conflict with his healthy survival.

MAN'S SYNTHETIC ENVIRONMENTS REFLECT HIS EMPHASIS ON QUANTITY, AND NEGLECT OF QUALITY

Man's epoch, among the other biotic specimens, is but a minute segment of the paleontological column representing the earth's populations of the many life forms as they have come and gone or remained.[6] His art of agriculture is not more than ten thousand years old. Again, and most significantly, the systematic development of science and its product, technology, began only some three centuries ago, while the neotechnical revolution, bringing mass production and the injection of energy into our system through the internal combustion engine, has taken place within our lifetime. In Paul Sears' words:

> Beginning with Darwin, biologists have shown the inseparability of life and its environment and the role of time in developing what has been, previous to man's advent, a beautifully adjusted dynamic equilibrium . . . A few hundred years of technology have already disrupted biological and geological processes that were established over eons . . . The present effect of technology is at variance with the pattern of conservation of material and energy that prevailed in the absence of man . . . There is nothing in untouched nature to compare with our extravagant use of energy and our failure to recycle essential materials . . . Danger lies in the disruption of the great dynamic processes which have made the earth so generally habitable. Included here are the water cycle, soil formation, and energy storage by plant life.[7]

Technology emphasizes quantity through mass production, high-speed chemical reactions, massive energy transfer to heat, and the shortened time required for output. The quality of the output depends on the particular parts and their order in the assembling operation. Manufacturing industry has brought much satisfaction and economic

---

[6] William A. Albrecht, "Wastebasket of the Earth: Man and His Habitat," *Bulletin of the Atomic Scientists,* Vol. 17 (1961), pp. 335–340.
[7] Paul B. Sears, "The Perspective of Time: Man and His Habitat," *ibid.,* Vol. 17 (1961), pp. 322–326.

success because its performance and products can be equated into values as monetary equivalents under sufficient control for perpetuation of productive capital, for matching production against consumption, and for guaranteeing profits. That satisfaction has generated the urban-based belief that agriculture, too, is only an industry, and should lend itself to corresponding increased output by means of increased power, higher speeds, and limited time according to the pattern of other industrial development.

That concept favors quantity output, but in agriculture it is slowly violating the quality of survival. Creation of living matter is not a process of assembling inert parts under human management. Rather, it is nature's unique integration of biochemical processes at low levels of energy and over extended periods of time. For that, man is at best a benign coöperator with the natural laws of growth and evolution, at worst an observer and collector in disregard of them.

Our disregard of ecology is perhaps the major reason for disruptions of the biotic phases of agriculture by technological manipulation. Plants and animals, both domestic and wild, have been disseminated by man almost to the limits of their environmental tolerances without regard for the quality of the soil. In some situations man's failure to manage the soil for more complete nutritional support of life forms has led to a decline in health. Environmental deficiencies, not readily recognized or remedied, have often been aggravated.

Selection of plant species and their propagation by mass production may sometimes result in metabolism modifications with increased output of carbohydrates and decrease of proteins. When the modified seed cannot be used for reproduction, the effects of the struggle for survival are eliminated. The plant's conversion of solar energy into carbohydrates is at an efficiency of about 30 per cent. But converted into proteins, the efficiency of energy conversion drops closer to 3 per cent. Modern industrial and economic concepts, based on quantitative yardsticks, do not tolerate such low conversion efficiencies. Hence the turn to hybrid corn, for example, and the consequent increase in quantity at the expense of quality. The mean of 10.3 per cent crude protein in corn has dropped to a low of half that during the last three or four decades in the United States. Corn is thereby started on the way toward self-extinction, becoming completely dependent on man for its survival.[8]

---

[8] Editor's Note: V. V. Rendig, Professor of Soil Chemistry at Davis, called attention in discussion to the relative "stability" of plant composition. Many properties of individual cells are of genetic origin and are not readily changed. Experiments which he cited, in which environment varied widely, have shown that many plant properties remain relatively constant in spite of environmental

Along with the emphasis on carbohydrates for quantity production of cereals, there has been a similar selection and propagation of farm animals for rapid gain in weight. Much of the natural struggle for survival is eliminated by early castration. Feed is mechanically and chemically manipulated for the purpose of tranquilizing and fattening the beasts in minimum time. In the process much of the quality needed for healthy survival is lost, and the life span is shortened to less than a year in cattle raised for "baby beef" and even less for "ton-litter" pigs. Innate resistance to disease seems to carry the animals in sufficient health during these brief periods to escape preslaughter death in spite of metabolic degeneration or microbial invasions.

Studies of diseases and their many symptoms, together with biochemical irregularities correlated with them, emphasize the absence of natural health. We have not yet sufficiently recognized the possibility that microbial invasions of crops and livestock may be due to the decreasing ability of domesticated plants and animals to maintain their molecular biology, through their own instinct-guided nutrition, at the high levels required for healthy coexistence with microbes.

The biological or ecological view for the maintenance and preservation of species has been replaced by a commercial attitude which advocates mass destruction of microbes. Forgotten is the biotic desirability of maintaining the natural self-protection by which plants

---

variables. One example given was the well-known ability of certain aquatic species to accumulate potassium even though the nutrient medium is much higher in sodium.

The ability of the plant to "defend" itself against its environment can be exceeded, however. If this were not true, "foliar diagnosis" as a means of evaluating differences in soil fertility would not be the useful tool that it is. The "luxury consumption" of an ion such as potassium was also mentioned as a case in point. These and other examples show that the concentration of a soil-derived nutrient in the tissue of a plant bears some relationship to the level of that nutrient in the soil. It was also pointed out that the different levels of nutrients in soils may be reflected in plants by virtue of their role in some physiological reaction in the plant. As shown by an example taken from his own work on sugar metabolism, plants are able to perform reactions which are not considered as part of normal metabolism. Abnormal nutrition could induce the plant to call upon these normally latent metabolic routes. A note of caution was interjected regarding the evaluation of any of these changes which may be induced by variations in soil fertility, in terms of crop quality. Whether any such change could be called an effect on "quality" will depend upon the kind of organism that consumes it. Thus Professor Rendig's experience has not shown that the more vigorous plant grown under conditions of favorable nutrition can defend itself better against preying insects than a less thrifty plant. In fact, the contrary has been observed in some kinds of nutritional deficiencies. Dr. Agnes Fay Morgan's experience with fox skins, one from a pantothenic acid-deficient animal and one from its normal litter mate illustrated this. The gray-furred deficient skin was destroyed by moths in storage, but the dark-furred normal skin remained intact.

and animals were so well adapted to their environment before domestication. Problems of degeneration, viewed only as pests and diseases, have been aggravated, with a parallel reduction in the quality of product for nutritional support of man in better health.

Another illustration of the disruption of a natural equilibrium is the increasing failure of choice legume crops. This is a concomitant of our emphasis on quantitative increase in nitrogen measured only through ash analyses and the acceptance of that monovalent element as the bona fide indicator of total protein. We have not taken cognizance of the simple, natural fact that an amino acid may contain two atoms of nitrogen: one of amino form and of regular digestive value, the other in the four-carbon-ring form. The latter is indigestible and excreted in that same structure; so the chemical tests falsify the truly protein quality by just 100 per cent.

Chemical tests, often promoted by the desire to sell fertilizers, are not the equal of biotic assays, yet they are the props for the large but nutritionally unbalanced yields of nonlegume plants fertilized with nitrogen which today's farmer is producing. At the same time, the animal's refusal to eat the grasses in the hummocks of pastures fertilized in spots by its own fecal and urinary droppings indicates the imbalance of the fertilizer used. This is an example of one of the many disruptions of the natural coexistence of plants and microbes in equilibrium when monovalent salts of high solubility and speedy chemical reactions are introduced into the soil, even by careful placement away from the planted seeds.

Our mass-production technology applied to agriculture has produced economic surpluses of carbohydrates in the form of cereals and of fats from livestock. Industrial uses of those two materials offer no economic relief, when alcohol from treatment of crude oil can displace natural fermentation of grains, and chemical detergents can substitute for soaps. But now that our tastes for fats as carriers of flavor have been easily satisfied to include energy needs, and we have recognized the cardiac implications of obesity, natural proteins and their related organic and inorganic substances are becoming recognized in household terms as requisites for health. The increasing preference for naturally fresh, but highly perishable, foods as opposed to those industrially synthesized and preserved for long shelf life has led us to speculate increasingly about the relationship between poor health, biochemical degeneration, and breakdown in molecular biology. It is compelling us to trace these back from ourselves to our livestock, to our crops, and finally to our deficient or mismanaged soil.

Quality in foods and feeds is at last under critical scrutiny for protein production through healthy species grown on fertile soils.

When "the proof of the pudding is in the eating," a naturally healthy survival is the proof of the quality in agricultural production. For the demonstration of that, biotic assays must supplement chemical tests. Quality in food and feed products for healthy existence, as the species itself reflects it, must prevail first for microbes and plants through the medium of the soil. Then it must prevail as interrelations among all the higher segments or trophic levels, if quality in production is to support man in good health as the top segment of that biotic arrangement.

Nutrition by natural choice is coming under the searchlight of science. We are beginning to measure the delicate impulses from the animal's taste buds, correlated with those of olfactory origin, for taste reactions are significant natural means of measuring quality by choice. Such choice, expressed in amounts consumed, may be the major means of putting quality of production under quantitative categorization by biotic means.

Animal discrimination in quality may be very delicately adjusted. A hog's selection of corn showed increasing consumption of that grain when it was organically fertilized with sweet clover. Rabbits have refused the same plant species when nitrogen was applied in increasing amounts, and the sexual vigor of the males was destroyed within a week when they were compelled to subsist on such plants.

Cattle are equally capable connoisseurs, or chemists. For eight successive years a herd selected one haystack, of four used as winter feed, in which the smallest share of its hay had been grown on soil treated with calcium cyanamid and superphosphate. Much more delicate discriminations by cattle recently have been observed now that inorganic salts may be offered cattle to assay the completeness of the ration, along with spectrographic analyses of the forages for trace-element content. In testing the opposing effects of copper and molybdenum, choices as delicate as ten parts per million have been claimed for dairy cattle. It might be questioned whether the choice of the castrated male would be as delicately adjusted.

The natural behavior of living species in relation to agricultural production suggests hope for prevention of degeneration of our crops, our livestock, and ourselves. We must pattern soil management for production that is in conformity with ecological knowledge and that does not go counter to biotic laws as revealed in evolution and the survival of species. The fragmentation of our thinking and action under technological and economic stimuli must be counterbalanced by integration into the geoclimatic foundation outlining the natural ecology. The species which we manage must be directed in more complete conformity with the factors we can assess, first, in the natural state and then duplicate in domestication. Fuller knowledge of the natural forces operating in agriculture, and those humbly modified by us with minimal disruption of natural processes, will result in quality first, to which quantity will be a sequel. Quality is a result of judgment by reasoned experiences, and hence comes only with maturity. Eventually we may arrive at maturity in managing soil resources more efficiently for the support of the entire biotic pyramid of which man is the apex and a dependent on all that is below him.

THE USE OF FIRE IN WILDLAND MANAGEMENT IN CALIFORNIA

Harold H. Biswell

Prescribed burning is the judicious use of fire for a constructive purpose and according to a definite plan carried out or supervised by the proper authorities. Done with care, it can be an efficient tool for wildland management. It can be used in selected places to reduce fuels and lessen the severity of wildfires, to manipulate forest stands toward more desirable species, to improve habitat for wildlife, to increase forage for livestock, and to maintain or improve landscape and scenic values.

A management plan for controlled burning will be concerned with such questions as: Why burn? Where to burn? When to burn? How to burn? The definition for prescribed burning given in the glossary of *Forestry Terminology* is as follows:

The skillful application of fire to natural fuels under conditions of weather, fuel moisture, soil moisture, etc., that will allow confinement of the fire to a predetermined area and at the same time will produce the intensity of heat and rate of spread required to accomplish certain planned benefits to one or more objectives of silviculture, wildlife management, grazing, hazard reduction, etc. Its objective is to employ fire scientifically to realize maximum net benefits at minimum damage and acceptable cost.

In wildland management, it is wise to study nature and fit management to nature's way. This may not always be possible under present-day conditions, but when management goes contrary to nature's way, difficult problems are often created such as those we face in widespread and severe forest fire hazards in California.

For the past fifteen years or so, I have studied fire ecology in three vegetation types in California: ponderosa pine, chamise chaparral, and woodland-grass. Each type requires a different prescription, and is discussed separately.

FIRE ECOLOGY IN PONDEROSA PINE

In this section four main points are considered. The first is concerned with primitive mountain forests and how they developed in nature with frequent light fires, which were an essential element in the balance of nature. The second topic concerns changes that have taken place in mountain forests that have led to extreme fire hazards as a result of fire protection. These changes are the result of an upset in the balance of nature or of fire suppression. The third topic is prescribed burning in ponderosa pine by means of light fires that are somewhat similar to those that occurred naturally before fire protection was started. Fourth, and perhaps most important, an idealized program for ponderosa pine management, designed to keep fire hazards low on a continuing basis, is suggested. In this plan, ponderosa pine is permitted to develop much as I believe it did in nature.

From such books as *Mountaineering in California* by Clarence King, *The Mountains of California* by John Muir, and *The Wilderness World of John Muir* by Edwin Teale, we can gain some idea of the appearance of our primitive forests. Descriptions by Clarence King and John Muir led me to believe that the forests originally were relatively clean, open, and park-like. As Clarence King wrote of going into the Sierra Nevada,

Passing from the glare of the open country into the dusky forest, one seems to enter a door, and ride into a vast covered hall. The whole sensation is of being roofed and enclosed. You are never tired of gazing down long vistas, where, in stately groups, stand tall shafts of pine. Columns they are, each with its own characteristic tinting and finish, yet all standing together with the air of relationship and harmony . . . Here and there are wide open spaces around which the trees group themselves in majestic ranks . . . Whenever the ground opened level before us, we gave our horses the reins, and went at a free gallop through the forest; the animals realized that they were going home and pressed forward with the greatest spirit. A good-sized log across our route seemed to be an object of special amusement to Kaweah, who seized the bits in his teeth and, dancing up, crouched, and cleared it with a mighty bound, in a manner that was indeed inspiring, yet left me with the impression that one was enough of that sort of thing.

In John Muir's book we find this statement:

The inviting openness of the Sierra woods is one of their most distinguishing characteristics. The trees of all of the species stand more or less apart in groves, or in small irregular groups, enabling one to find a way nearly everywhere, along sunny colonnades and through openings that have a smooth, park-like surface, strewn with brown needles and burrs . . . One would experience but little difficulty in riding on horseback through the successive belts all the way up to the storm-beaten fringes of the icy peaks.

What kept these forests clean, open, and park-like under natural conditions? Probably the most important agent was fire. Because fires were frequent, they were light, even though many of them occurred at the driest time of the year.

John Muir referred to fire as the master controller of the distribution of trees, and described a fire as it entered a grove of Giant Sequoia trees between the Middle and East forks of the Kaweah, in early September, the driest time of the year:

The fire came racing up the steep chaparral-covered slopes of the East Fork canyon with passionate enthusiasm in a broad cataract of flames. . . . But as soon as the deep forest was reached, the ungovernable flood became calm like a torrent entering a lake, creeping and spreading beneath the trees. . . . There was no danger of being chased and hemmed in, for in the main forest belt of the Sierra, even when swift winds are blowing, fires seldom or never sweep over the trees in broad all-embracing sheets as they do in the dense Rocky Mountain woods and in those of the Cascade Mountains of Oregon and Washington. Here they creep from tree to tree with tranquil deliberation, allowing close observation.

According to this statement, John Muir did not know of a single crown fire occurring in the Sierra Nevada.

By dating the fire scars on the trees, plant ecologists have been able to tell a great deal about the fires that burned hundreds of years ago. In a study of dry rot of incense cedar, evidence showed that fires produced many scars throughout the entire pine-forest region, at intervals of three to eleven years, during the late seventeenth century and throughout the eighteenth and nineteenth centuries.

A study in the Stanislaus National Forest, where detailed data were collected on 74 acres, showed that 221 distinct fires swept that area between 1454 and 1912. This would average one fire about every two years. I believe that those fires were set both by lightning and by Indians.

In view of the large number of natural fires in primitive times, I decided to check the number and distribution of lightning fires occurring today in the Sierra Nevada. It seemed reasonable to assume that the pattern would be somewhat similar to that of very early times, the main difference being that lightning fires are now quickly suppressed. For the study I selected a township at the Tuolumne Grove of Big Trees and another near the Mariposa Grove, both in Yosemite National Park. In the township at Tuolumne, 39 lightning fires were suppressed in the nine years from 1951 to 1959, inclusive; for the Mariposa Grove the number was 36. Similar data were obtained for a township at Sloat, in the mountains east of Chico, and for another at Pinecrest, where some of the fire-scar data on incense cedar were taken. In the first instance, 24 fires were suppressed; in the latter, 18. Lightning fires were recorded for all townships in every year except 1954. Similar results were obtained for the Plumas, Stanislaus, Sierra, and Sequoia national forests in the Sierra Nevada.

These data lead to only one conclusion: in aboriginal times, fires must have been widespread in the Sierra Nevada nearly every year, not every other year or every eight years, as indicated by the fire-scar data on incense cedar. This does not mean, of course, that every spot burned every year. However, since enough needles drop each year beneath ponderosa pine to carry fire on the ground, perhaps as many as 6,000,000 acres of forest lands (of the approximately 18,000,0J0) in California burned every year. But the fires were very light, merely creeping about, and were necessary for the development of the primeval forest. These fires in no way resembled the crown fires of today.

Besides the fires started by lightning, others were set by Indians in aboriginal times. This is borne out by explorers and naturalists who observed the results at first hand, such as Galen Clark, who for many years was guardian of Yosemite; Dr. L. H. Bunnell, a member of the 1851 Yosemite discovery party; and Joaquin Miller, who wrote in 1887: "In the spring . . . the old squaws began to look about for the little dry spots of headland or sunny valley and as fast as dry spots appeared, they would be burned. In this way the fire was always the servant, never the master. . . . By this means, the Indians always kept their forests open, pure and fruitful and conflagrations were unknown." From this evidence it seems clear that the Sierra Nevada forests were originally clean, open, and park-like and that the most important agent in maintaining this condition was fire, frequent and light. In aboriginal times, sweeping crown fires were unknown.

When forest management was initiated in California around the turn of the century, one of its objectives was adequate fire protection, the first essential in good management. As a result of protection, however, fire hazards have increased in many places and have made wildfire suppression very costly and difficult. Even though we have the best fire suppression forces in the world, and excellent work is being done, some of the fires become big. In the past few years they have destroyed a wealth of vital resources.

In the early fall of 1959, in one 36-hour period, 50,000 acres were charred in the Sierra Nevada; the final estimates of damage ran as high as $66,000,000 in timber and watershed alone. This fire was started from debris burned during highway construction. Wildfires of this sort are a major disaster. But as the population and wealth of California increase, the damage from wildfires and the cost of suppression will become even greater. The contrast of affairs at present with those in aboriginal times points up the dilemma facing those concerned with managing our wildland resources.

As a result of fire protection, two extremely hazardous conditions have developed in the Sierra Nevada forests. First, shade-tolerant white fir and incense cedar have formed dense thickets in many places in the understory of larger trees. Many people regard this growth of understory as a natural, undisturbed succession toward climax vegetation. They fail to recognize that in earlier times frequent small fires were a natural element in the environment, and were effective in suppressing the hazardous growth of understory trees.

The second condition is the great increase in debris on the forest floor. Many forests that were relatively clean, open, and park-like in earlier times, when frequent small fires kept the undergrowth cleared, are now so full of dead material and young trees and brush as to be nearly impassable. (See pl. 1.) Decay of dead material has not kept up with the rate of accumulation, because summers are too dry and winters too cold for the necessary bacterial activity. As a result of these two changes in the Sierra Nevada forests, brought about by fire control unaccompanied by compensatory controlled burning or by other means, we now have a tiger by the tail, so to speak. The great forest fires are hazardous, but it is dangerous also to use light ground fires, simulating those in nature, to reduce fuels and try to restore the original forest conditions. To reclaim these areas we must do a great deal of cutting by hand, and piling and burning during the wet season, in order to make prescribed burning possible later. This, of course, will be very expensive, but it must be done at any cost if we expect to save our forests from destruction by wildfires.

Back in the late forties, after I had studied the ecology of aboriginal pine forests and observed the present high fire hazard conditions, it became obvious to me that new ways must be sought to modify forest fuels, and new practices developed to avert the mounting danger and costs of suppressing great wildfires. Since light fires were so effective in nature in creating and maintaining clean, open, and park-like conditions, I thought it worth while to do some research on this phenomenon. My associates and I started experiments in prescribed burning on the Teaford Forest in the central Sierra Nevada near North Fork in April, 1951, and at Hoberg's in the North Coast Range, in the fall of that year. (I had just finished six years of study on the use of fire in the pine forests of the southeastern United States and was favorably impressed with the results. In that area, perhaps a million or more acres of pine lands are prescribed-burned each year.)

In nature, ponderosa pine forms a fire subclimax forest type, and it is possible that frequent light fires are necessary for its best development. It was already known that ponderosa pine is thick-barked and relatively fire-resistant.

Why prescribe-burn?—The objective is to reduce debris on the forest floor and kill brush and small trees in the understory of larger pines, thereby reducing fire hazards and the danger of wildfires. (See pl. 2, *a* and *b*.)

Where to burn?—In ponderosa pine prescribed burning can be done where needles on the ground provide sufficient fuel to carry the ground fire. This cannot be done by the heavier debris or by brush or trees. There should be enough trees above ten to twelve feet in height for a stand. In areas where the debris is relatively light, however, burning can be done under trees that are only six to eight feet tall, if there are enough pine needles to carry the fire.

When to burn?—Burn when the trees are dormant, from November through March—perhaps a little earlier or a little later in some years, depending on moisture and weather. Prescribed burning should be done only after enough rain has fallen to wet the duff to the mineral soil so that the soil is moist or wet in the areas of deepest duff. To achieve this in heavy duff may require three inches or so of rain, but in light duffs, perhaps after one or two burns, an inch or so of rain will be sufficient. After the heavy fuels have been removed and the small trees destroyed, prescribed burning can be done later in the spring or perhaps into the summer, but not until after three or four burns.

In heavy fuels, and especially for the first burn, the air temperature should be less than 60°F and the humidity 35 per cent or higher. After one burn, the requirement for humidity can be lowered to 25 per cent or so, and that for temperature raised to 75°F. Cool air causes heat to dissipate rapidly, and keeps scorching to a low level. Start burning two or three days after rain or when the upper pine needles are dry enough to carry fire. If the fuel becomes very dry and seems to burn vigorously, wait until after another rain before setting more fire. At Hoberg's, burning could be done for 47 to 74 days each year over a five-year period. Wind forecasts should be checked. It is permissible to burn during a light breeze or wind, but heavy wind should be avoided. One of the greatest dangers in burning is a rising wind, which sometimes comes up suddenly.

How to burn?—Start prescribed burning on the top of a ridge. Rake the needles toward the area to be burned for three or four feet and set the fire on the *edge* of the cleared line. Do this slowly, and watch the fire at all times. If the fire burns satisfactorily and does not seem difficult to control, the line can be lengthened. Tend the fire so that it always moves downhill. A fire that moves gently downhill may burn out of control if permitted to burn uphill. After a wide fire lane has burned out, strip burning may be desirable. This means carrying the fire downslope a short distance (15 to 20 feet) and letting it burn uphill. This speeds up the operation and will give more fire and heat in certain places.

Obtain a fire permit, and notify neighbors about burning. Because burning should be done only when conditions are right, the date cannot be set far in advance. A back pump or two should be near at hand. If large standby crews and equipment might be needed, however, wait until the conditions are not so extreme. The fire will move downhill through pine needles at a rate of less than one foot per minute. Therefore, a large acreage cannot be covered in a short time with a short line. Fires set at different times should be carefully planned for greatest safety. After a prescribed burning program is well under way, it becomes possible to burn uphill, and the area can then be covered very rapidly.

In some areas it may be desirable to follow prescribed burning with cleanup work. This consists of piling and burning dead brush, slash, and dead fallen trees. Start a fire and pile up dead material gradually as it burns. A good worker can keep half a dozen piles going at one time, and bulldozers can be used in open areas to advantage. This kind of burning can be done under a wider range of weather conditions than can prescribed burning, because the fire hazard has already been reduced; also, techniques have been developed for starting fires under wet conditions, or even in light rain.

Who should burn?—Prescribed burning should be done by men with training and experience. Intelligence, good judgment, and patience are required also. Being a professional range man or a professional forester does not automatically qualify one to do prescribed burning. Even an expert in fighting wildfires may not have the attributes and training necessary to do prescribed burning.

After an area has been prescribe-burned and cleaned, the falling pine needles can drop on the ground rather than drape themselves over the debris and dead brush; consequently, the fire hazard builds back very slowly and the benefits from burning can last for many years. Cleanup burning is rather expensive, but when the cost is divided by the years of benefit the operation appears to be very cheap indeed.

Whereas prescribed burning is looked upon as a skilled technical job, cleanup burning is mostly manual labor, and possibly the operating of special equipment.

The danger of wildfires and the high cost of suppression have become so critical that we should give greater attention to forest management practices that will assure lower fire hazards on a continuing basis. For this purpose we have formulated an idealized management program for ponderosa pine which results in an overall, uneven-aged forest made up of even-aged groups of trees, about as they developed in nature. The program provides for (1) clear cutting in spots, at the end of a rotation cycle; (2) cleaning and prompt restocking of those spots; (3) the exclusion of prescribed fire in the spots of reproduction until the new trees are ten to twelve feet tall, or until enough pine needles accumulate on the ground to carry a light ground fire; (4) periodic burning thereafter to keep out the understory reproduction of trees and brush; (5) frequent intermediate harvest cuts; and (6) cleaning of all slash after each harvesting. Pines should not be permitted to give way exclusively to other species, for pine needles are excellent fuel to carry ground fires in prescribed burning. (See pl. 3.)

To visualize this operation more clearly, imagine that you have 3,000 acres, that markets are available for all trees over six inches DBH (diameter breast height), and that you want to harvest trees every year. When trees are in properly stocked stands, they grow fast, and light, intermediate harvest cuts can be made frequently, perhaps every ten years. At that rate, and with the rotation cycle stretched to 150 years, 20 acres would be clear-cut each year, perhaps in ten or fifteen different spots, and 200 acres would be thinned in intermediate harvest cuts. The clear-cut spots would be cleaned of slash and managed to obtain fully stocked stands of reproduction, either by natural seeding or planting. Prescribed fire must be excluded from all new reproduction until the trees are ten to twelve feet tall, or until there are sufficient pine needles on the ground to carry fire. After that time, burning can be used to maintain conditions for low fire hazards. No more tree reproduction is wanted until the end of the rotation cycle, because reproduction under older trees adds greatly to the forest fire hazard. This idealized management is nothing more than farming crops of trees; you get a stand of trees after clear-cutting on small spots at the end of the rotation cycle, and then you are concerned for many years with the protection and growing out and harvesting of that crop. (See pl. 4.)

As I visualize this plan, it is almost identical with the way in which the forest reproduced itself and developed when nature was free to operate in its own fashion. During aboriginal times, the forest cover and light fires must have been in nearly perfect balance—nature's balance. From this research I have concluded that prescribed burning in ponderosa pine simulates the light fires that functioned in nature to develop a stable type of fire subclimax forest. Thus in the judicious use of fire we are imitating nature.

No wildfires have occurred on the Teaford Forest since the prescribed burning was started in 1951. At Hoberg's, however, where four fires have occurred, the prescribed burning greatly reduced the wildfire hazard. One fire provided a nearly perfect demonstration. In late August, 1962, a severe wildfire burned into Hoberg's and into an area that had been prescribe-burned. On the outside, the fire had been crowning, and had burned off the needles and killed nearly every tree in its path. But when the fire reached the area that had been prescribe-burned, it went to the ground, became relatively calm, and progressed gently through the pine needles. The fire acted much like one in prescribed burning, and it may have done more good than harm by removing debris and further reducing the danger of wildfires. In the fall, when the rains started, this area was well covered with a new crop of pine needles, and it was difficult to see that it had been burned. On the outside, however, the soil was nearly bare, and surface runoff was heavy. There the pine needles had burned on the trees; so none remained to fall and cover the soil. Such runoff can cause erosion on the slopes and flooding below. The other three wildfires at Hoberg's were minor, and burned only through pine needles on the ground. They did not scorch any needles on the trees.

The results here are very much in accord with studies reported by Harold Weaver, who worked in ponderosa pine in Washington and Arizona. He reported that prescribed burning reduced fire incidence by 78 per cent, damage by 87 per cent, and cost of control by 55 per cent.

FIRE ECOLOGY IN CHAMISE CHAPARRAL

Chamise chaparral predominates on about 7,300,000 acres of wildlands in California. Other species commonly associated with chamise are interior liveoak, buckbrushes, manzanitas, toyon, and chaparral pea. The shrubs form an almost closed canopy, so complete as to preclude much herbaceous vegetation as understory. Most of the shrubs crown-sprout after fire.

Chamise brushlands have been looked upon as valuable chiefly for game and watershed. The soils are usually shallow, although in a few areas they are deep enough to be cleared for agricultural use. Some lands are not steep, but others are rugged and isolated and are not well suited for prescribed burning. Wildfires are frequent and widespread, and in recent years some have been particularly bad in denuding watersheds and destroying homes in or near chaparral.

Chamise chaparral was studied in the north coast range in Lake County for several years in connection with prescribed burning for game range improvement. (See pl. 5.) The brushlands there form two cover types: one is predominantly chamise; the other contains a mixture of broad-leaf shrubs and trees with chamise, known as mixed chaparral. The first occurs mainly on south-facing slopes and drier sites, while the mixed chaparral is found on the more mesic, north-facing exposures and in ravines.

Why burn?—The objective in prescribed burning is to improve habitat conditions for game by reducing the brush cover in spots to create openings, "edge," and sprouting, and to introduce palatable herbaceous species for deer use in winter and early spring. Deliberate burning of the brush in spots, leaving 25 per cent or so unburned, results in a favorable interspersion of browse, grasses and forbs, and protective cover. It was already known that a year-long source of water was available in the ravines.

Where to burn?—Prescribed burning is most needed in mature chaparral where few sprouts and little herbaceous forage grow for wildlife during the winter and spring months. A mixed type of shrub cover provides greater variety and better browse than does a nearly pure chamise cover.

When to burn?—For game range improvement, either spring or late fall burning is satisfactory in chamise brushlands. When chamise areas are surrounded by grasslands, spring burning is relatively safe from the standpoint of fire containment because the brush will burn when the surrounding grass is green and will not carry fire.

How to burn?—The method of burning depends on the continuity of brush cover and the surrounding vegetation. In areas where brush is surrounded by grass, spots of chamise brush can be burned in early spring with a minimum of firebreak preparation. Prescribed burning is done by setting a fire at the base of a slope and letting the fire burn uphill to the ridgetop. Under proper conditions the fire will not spread to the sides and will go out on top of the ridge. (See pl. 6.) If prescribed burning is done in summer or fall, elaborate firebreaks may be necessary for adequate control.

Burned spots for game should usually be small in order to form as much edge as possible. The acreage to be burned should be decided upon before burning is started. In general, deer populations in areas of heavy brush are fairly low, and the burns should be kept small. Scattered spots of five to ten acres each are probably sufficient for initiating a program of managing chamise brushlands. In two or three years a larger area could be prescribe-burned, especially where deer use has been heavy. If the deer population increases as the treated area increases, it might be desirable to go rather fast in cover manipulation. In the end, perhaps as much as 75 per cent of an area can be treated and 25 per cent left as heavy brush for cover.

An area of prescribe-burned chamise chaparral was compared with a similar unburned area as a control. Counts of deer in the burned area showed a summer population density of about 98 per square mile after the initial burning treatment. This rose to 131 in the second year, and dropped to 84 in the fifth and sixth years. In the dense, untreated brush the summer density was only 30 deer per square mile. Ovulation rate in adult deer was 175 per cent in treated brush and only 82 per cent in untreated brush. Deer weights were higher in prescribe-burned brush than in the untreated area. (See pl. 7.)

Studies have not gone far enough to determine precisely whether most of the burning for game should be done in spring or in fall. After spring burning, new sprouts appear in three or four weeks and supply a highly palatable, protein-rich browse for deer during the dry summer months. However, after burns made in April, new seedlings do not appear until the next year, and by that time seedbed conditions are not favorable for them. The greatest number of new seedlings come in the spring after fall burning. Apparently spring burning favors sprouting shrubs over nonsprouting species. Since some of the better shrubs are nonsprouting species, perhaps a combination of spring and fall burning is preferable to burning only in spring. The frequency of prescribed burning is important also in vegetation manipulation. Frequent fires favor sprouting species, especially where browsing retards those that do not reproduce by sprouting. When nonsprouting shrubs are present, a second fire may eliminate this species if it has not produced seed between the two fires. This would be unfavorable for the supply of browse for deer.

FIRE ECOLOGY IN WOODLAND-GRASS VEGETATION

Woodland-grass vegetation predominates on about 7,500,000 acres in California. It is used primarily for livestock grazing. Since brush has invaded or increased over much of this grazing area, landowners are demanding prescribed burning to reduce woody vegetation and improve forage for livestock. (See pl. 8.) The principal woody species are buckbrush, manzanita, digger pine, and oak. In this association, a large percentage of the brush cover is made up of nonsprouting species. The soils are fairly deep, and productive of grasses.

Why burn?—The primary objective in prescribed burning in woodland-grass is to kill some of the woody vegetation and replace it with grasses to increase the grazing capacity for domestic livestock. Usually, however, the brush cover is not completely destroyed but only thinned; so conditions are improved for deer and upland game birds also. (See pl. 9.) In some places, spring flow increases after brush replacement, and the increased water is beneficial to both livestock and wildlife. Another very important benefit is a reduction in wildfire hazards.

Where to burn?—Prescribed burning is most successful in areas in which a large percentage of the brush cover is made up of nonsprouting species. In interior liveoak it does little good, because this species sprouts vigorously. In dense blue oak, fire that is burning grass in the understory may not be intense enough to kill many of the trees.

When to burn?—Prescribed burning in woodland-grass vegetation is usually done in the summer when the grasses are dry and will carry fire from one brush area to another. If areas of dense brush have been pushed over with a bulldozer, these may be prescribe-burned in the spring or early fall when the fire hazards in surrounding areas are relatively low.

How to burn?—Because prescribed burning in woodland-grass vegetation is done during the dry summertime, elaborate fire lanes are needed to ensure that the fire will be contained in the prescribed area. Even then, about 11 per cent of the prescribed fires escape over the control lines. Usually ranchers, working with farm advisors and State Division of Forestry personnel, prescribe how the burns are to be carried out, and later inspect the preparations for containing the fire. The burning is done by groups of ranchers, perhaps fifty or sixty on each fire, with both ignition and control equipment. The fire boss is a rancher with much experience in prescribed burning. The area covered in any one burn may be 1,000 or more acres, but, regardless of size, the objective should be to start and complete a burn in one day. Both experience and economic studies have shown that 400 to 600 acres is a convenient size to burn in one day. The fire is started about 8:00 A.M. on the side against the prevailing winds, and the fire setters work in both directions around the area, to join forces about 2:00 P.M. The fire can then burn across the area with enough intensity to destroy much of the brush. (See pl. 10.)

Ranchers sometimes carry out extensive preburn treatments, such as pushing the brush over with a bulldozer blade held about six inches above the soil, to create dry fuel for better combustion. Although some of the prescribed burns escape even in suitable weather, ranchers usually do not burn during critical fire periods when wildfires are most difficult to control.

A second prescribed burn is almost always needed for each area in order to destroy brush seedlings that come after the first fire. Fire enhances germination of the seed of many species by cracking hard seed coats, and it also prepares an ideal seedbed for seedlings. Therefore, two burns in succession are necessary: the first, to kill the old plants so that no more seedlings are produced; and a second, to kill the seedlings that appear after the first fire. With two such burns, the shrubby cover can be suppressed for many years.

The amount of increase in grazing capacity for domestic livestock depends upon the initial density of the brush cover, success in removal, and the forage-producing capacity of the soil after the brush has been destroyed. In some areas prescribed burning has greatly increased the grazing capacity, but in other places the increase has been minor. Prescribed burning is not the only treatment in woodland-grass range improvement. Chemicals and mechanical equipment are used also, and areas prescribe-burned are reseeded and managed for maximum production. Spring flow sometimes increases after prescribed burning, with more water available to both livestock and game.

SUMMARY

Research and experience indicate that prescribed burning, done with care, can be a useful tool in wildland management, to reduce fire hazards, improve habitat conditions for wildlife, and improve grazing for livestock.

REFERENCES

Arnold, Joseph F.
  1963. *Uses of Fire in the Management of Arizona Watersheds.* Second
    Tall Timbers Fire Ecology Conference. Tall Timbers Research
    Station, Tallahassee, Florida. Pp. 99–111.
Biswell, H. H.
  1959. "Man and Fire in Ponderosa Pine in the Sierra Nevada of Cali-
    fornia," *Sierra Club Bulletin,* Vol. 44(7), pp. 44–53.
  1961. "The Big Trees and Fire," *National Parks Magazine,* April.
  1963. *Research in Wildland Fire Ecology in California.* Second Tall
    Timbers Fire Ecology Conference. Pp. 63–98.
Harper, Roland M.
  1962. *Historical Notes on the Relation of Fires to Forests.* First Tall
    Timbers Fire Ecology Conference. Tall Timbers Research Sta-
    tion, Tallahassee, Florida. Pp. 11–29.
Humphrey, Robert R.
  1962. *Fire as a Factor in Range Ecology.* New York: The Ronald
    Press Co. Pp. 148–189.
Jepson, W. L.
  1921. "The Fire-Type Forest of the Sierra Nevada," *Intercollegiate
    Forestry Club Annual,* Vol. 1(4), pp. 7–10.
King, Clarence
  1871. *Mountaineering in California.* Reprinted with preface and notes
    by Francis P. Farquhar. Copyright 1935.
Klemmedson, J. O., A. M. Schultz, H. Jenny, and H. H. Biswell
  1962. "Effect of Prescribed Burning of Forest Litter on Total Soil
    Nitrogen," *Soil Science Society of America, Proceedings,* Vol.
    26(2), pp. 200–202.
Komarek, E. V., Sr.
  1962. *Fire Ecology.* First Tall Timbers Fire Ecology Conference. Pp.
    95–107.
  1964. *The Natural History of Lightning.* Third Tall Timbers Fire Ecol-
    ogy Conference. Tall Timbers Research Station, Tallahassee,
    Florida. Pp. 139–183.
Komarek, Roy
  1963. *Fire and the Changing Wildfire Habitat.* Second Tall Timbers
    Fire Ecology Conference. Pp. 35–43.
Kotok, E. I.
  1934. "Fire, a Major Ecological Factor in the Pine Region of Cali-
    fornia," *Fifth Pacific Science Congress, Canada, Proceedings.*
    University of Toronto Press.
Leopold, A. S., S. A. Cain, C. M. Cottam, I. N. Gabrielson, and T. L.
    Kimball
  1963. "Wildlife Management in the National Parks." Report of Ad-

visory Board on Wildlife Management to Secretary of the Interior Udall. 23 pp. mimeo.

Lotti, Thomas
1962. *The Use of Prescribed Fire in the Silviculture of Loblolly Pine.* First Tall Timbers Fire Ecology Conference. Pp. 109–119.

Mason, Herbert L.
1955. "Do We Want Sugar Pine?" *Sierra Club Bulletin,* Vol. 40, pp. 40–44.

Miller, Howard A.
1963. *Use of Fire in Wildlife Management.* Second Tall Timbers Fire Ecology Conference. Pp. 19–30.

Muir, John
1894. *The Mountains of California.* New York: American Museum of Natural History, 1961.

Reynolds, Richard
1959. "Effects upon the Forest of Natural Fire and Aboriginal Burning in the Sierra Nevada." Master's thesis, Department of Geography, University of California, Berkeley.

Riebold, R. J.
1964. *Large-Scale Prescribed Burning.* Third Tall Timbers Fire Ecology Conference. Pp. 131–138.

Sauer, Carl O.
1950. "Grassland Climax, Fire, and Man," *Journal of Range Management,* Vol. 3, pp. 16–21.

Squires, John W.
1964. *Burning on Private Lands in Mississippi.* Third Tall Timbers Fire Ecology Conference. Pp. 1–9.

Stewart, Omer C.
1956. "Fire as the First Great Force Employed by Man," reprinted from *Man's Role in Changing the Face of the Earth.* Chicago: University of Chicago Press. Pp. 115–133.
1963. *Barriers to Understanding the Influence of Use of Fire by Aborigines on Vegetation.* Second Tall Timbers Fire Ecology Conference. Pp. 117–126.

Stoddard, H. L., Sr.
1962. *Use of Fire in Pine Forests and Game Lands of the Deep Southeast.* First Tall Timbers Fire Ecology Conference. Pp. 31–41.
1962. *Some Techniques of Controlled Burning in the Deep Southeast.* First Tall Timbers Fire Ecology Conference. Pp. 133–144.
1963. *Bird Habitat and Fire.* Second Tall Timbers Fire Ecology Conference. Pp. 163–176.

Sweeney, James R.
1956. "Responses of Vegetation to Fire," *University of California Publications in Botany,* Vol. 28(4), pp. 143–250.

Sweeney, James R., and H. H. Biswell
1961. "Quantitative Studies of the Removal of Litter and Duff by Fire

under Controlled Conditions," *Ecology,* Vol. 42(3), pp. 572–575.

Teale, Edwin Way
1954. *The Wilderness World of John Muir.* Boston: Houghton Mifflin Co.

U.S. Forest Service
1955. *California Aflame.* San Francisco, California.

Weaver, Harold
1955. "Fire as an Enemy, Friend, and Tool in Forest Management," *Journal of Forestry,* Vol. 53, pp. 499–504.
1957. "Effects of Prescribed Burning in Ponderosa Pine," *ibid.* Vol. 55, pp. 133–138.
1964. *Fire and Management Problems in Ponderosa Pine.* Third Tall Timbers Fire Ecology Conference. Pp. 61–79.

PLATE 1. Dead brush and other debris in understory of ponderosa pine present extremely high fire danger.

PLATE 2. Before (above) and after (below) prescribed burning in ponderosa pine. Flash fuel was removed and total weight of dead material was reduced by 51 per cent.

PLATE 3. Low-intensity fires in prescribed burning remove fine materials and much debris without appreciable damage to crop trees.

PLATE 4. This area was prescribe-burned in November, 1956. The remaining dead material was burned in numerous small piles. The fire hazard is low.

PLATE 5. Typical dense chamise brushland in Lake County, California. Such brushlands can be prescribe-burned to improve habitat conditions for blacktail deer.

PLATE 6. Prescribed burning upslope in chamise chaparral in May.

PLATE 7. After prescribed burning, this area was reseeded to grasses.
The intermixture of browse and grasses with unburned spots for cover
provides ideal habitat conditions for deer.

PLATE 8 *(left)*. In woodland-grass vegetation, brush can become so dense that little forage is available for livestock. The shrubs are too dense also for optimum habitat for game birds and deer. PLATE 9 *(right)*. Same as plate 8, but after prescribed burning with two fires.

PLATE 10. Prescribed burning in woodland-grass vegetation.

PROBLEMS OF QUALITY AND QUANTITY IN THE MANAGEMENT OF THE LIVING RESOURCES OF THE SEA   *Milner B. Schaefer*

The biota of the sea which man uses directly consists mostly of the animals up the food chain from the plants. Along the margin of the sea the large attached algae, such as the kelp off the California coast, may be harvested directly, and higher plants, such as eelgrass, are also taken from the marginal sea. However, most of the plants of the sea are microscopic phytoplankton, not amenable to direct harvesting because of their dispersion, rapid turnover, and consequent very small standing crop per unit volume of sea water. Many of the animals of the open sea, that is, the small zooplankton, are likewise not suited to direct harvesting. Therefore, unlike the land harvest, that of the sea consists mostly of carnivores, together with some of the herbivores. The sea, thus, is a poor source of carbohydrates, but a good source of animal protein and some fats.

The primary use of the living resources of the sea is for protein food for direct human consumption. However, a substantial and increasing share of the fishery harvest, amounting to 27 per cent in 1962, is used for the manufacture of fish meal and oil. Virtually all the fish meal is used as a protein supplement in feeds for poultry, cattle, and swine. Other industrial products from the sea include the alginates from seaweeds and some vitamin and amino acid preparations from fish; certain antibiotics may have a potential as commercial products. Fish oils and whale oils are used in the manufacture of margarine, and in soaps, paints, varnishes, and other industrial products.

Particularly important in the United States, and especially in California, is the use of the living resources of the sea for recreation. The pursuit of game fish species, such as tunas, marlin, yellowtail, bonito, flounders, and shore fishes is enjoyed by a large number of our citizens and by visitors from other states. A not negligible use of the sea's living resources is for the pleasure people get from observing the larger animals in their natural habitat. "Whale watching" has become very popular in the vicinity of San Diego during the migration of the gray whale to its breeding grounds in Baja California.

A crude measurement of quantity of the harvest of the sea is provided by the weight of the landed catch. This is an appropriate measure of food value also, for, as my colleagues in the fishery technology inform me, there is little variation in the protein value of fishes of different kinds. It is not so good a measure of nutritional value if we include the fats, because these vary from species to species and from one season to another, as well as with the age of the fish.

The measurement of quality is a great deal more difficult, since quality depends not only on the inherent properties of an object but also on the attitude of the user. The same kind of fish, or fish products, may be regarded as of very high quality or of very low quality in different groups, or even by different individuals within the same social group. For example, the common mussel is highly esteemed as a luxury food in much of Europe, and by a few people in California; but most people in California regard it as a very low-quality food and refuse to eat it. For organisms harvested commercially for human food, price is probably the most useful comparative measure of quality; the balance between cost of harvesting and human demand is expressed in the price which the consumer is willing to pay for the product. The most desirable species fetch the higher prices: oysters, shrimp, lobster, halibut, salmon, and tuna bring relatively high prices per pound, while species of equal nutritive value but held in lesser esteem, such as sardines, mackerel, cod, and hake, bring much lower prices. Indeed, the market for hake for human food is negligible in the United States, although it is widely used in Europe and in some Latin American countries. The quality of fish as a source of fish meal and oil can be measured by the oil content, and by the protein content of the resulting fish meal. These, again, are reflected in the comparative prices of the raw material, since there is little variation in the final product regardless of the source of the raw material.

Recreational uses depend very greatly on quality, but the quality of a recreational resource is almost impossible to measure. Obviously, the value of a species of fish to sportsfishermen is seldom related to its food value. The marlin is a highly prized sportsfish, but most of the recreational catch is not eaten. The quality of a species for sportsfishing depends partly upon the behavior of the fish when hooked;

sportsfishermen prefer species which are very active on the end of a line. It depends also on the abundance and ease of capture; if the fish are extremely scarce or seldom take a lure they are not highly regarded, but if they are abundant and easy to catch they are, again, not very highly prized. There are also wide individual preferences: the dedicated salmon fisherman, or striped bass fisherman, often has a low opinion of people who go fishing for flounders or rock fish, while the devotee who pursues swordfish, marlin, and the larger tunas regards much less highly the pursuit of barracuda, yellowtail, or kelp bass. It is thus probably useless to attempt to categorize the recreational resources in regard to quality. We need to maintain a wide diversity of sportsfish to accommodate a wide variety of tastes and personal abilities.

INTERNATIONAL CONSIDERATIONS

The management of the living resources of the sea, even when dealing with local problems concerning commercial or recreational fisheries, should be considered in a broad, world-wide context for five reasons:

1. Many of the fish species captured along the margin of the sea are highly migratory and thus are not under the sole jurisdiction of the coastal state, even though they are captured only by its citizens in the territorial sea. The salmon originating in the rivers of North America go far out to sea; some races migrate nearly to the Asiatic mainland. The albacore tuna and bluefin tuna which occur off the coast of California migrate clear across the North Pacific Ocean to the coasts of Japan. The population of jackmackerel which we fish off California extends at least halfway across the North Pacific Ocean. The yellowtail, sardines, anchovies, and other species, which are more narrowly confined, yet extend far beyond our territorial sea, also migrate to waters lying off Mexico.

2. Fish occurring in the high seas, beyond the margin of the narrow territorial sea, must be managed in accordance with the provisions of international law, which were codified in the Convention on Fishing and Conservation of the Living Resources of the High Seas negotiated at Geneva in 1958. Although this Convention has not yet come into force, it provides the framework within which we must conduct our management of these resources. The Convention provides that all states have the right for their nationals to engage in fishing on the high seas, but it also provides that all states have the duty to adopt, or to coöperate with other states in adopting, such measures for their respective nationals as may be necessary for the conserva-

tion of the living resources of the high seas. Article 2 provides the following definition: "As employed in this Convention, the expression 'conservation of the living resources of the high seas' means the aggregate of the measures rendering possible the optimum sustainable yield from those resources so as to secure a maximum supply of food and other marine products. Conservation programs should be formulated with a view to securing in the first place a supply of food for human consumption." The quantitative aspects of the harvest receive the primary emphasis, for the very good reason that, at the present stage of history, it is impossible to reach agreement on the criteria of quality. Several economists have criticized the Convention for stating that the objective of conservation is to secure a maximum supply of food and other marine products, since, in their view, a lesser quantity, which could be taken at much lower cost, would provide a larger net economic yield. Here, again, this is theoretically a sound objection, but where we are dealing with a diversity of nations placing quite different evaluations on marine products, and having quite different costs of harvesting them, it is not feasible to establish any criteria other than the maximum sustainable harvest, which is a natural property of fish stocks. Where a fish stock is not being fully utilized, no need for conservation measures can be asserted. One may not, internationally, limit the harvest of an unutilized resource. If there are resources lying off our shores which we are not fully utilizing, we cannot deny their use to others. Even where the resource is being fully utilized, we must recognize the rights of other nations to participate in the harvest on an equitable basis. The Convention spells out in detail the methods by which conservation regulations in the high seas are to be established, and for adjudication of disputes among nations.

No provision is made in this Convention for preferential use of any resource for recreation.

3. It has been estimated that some 500,000,000 of the world's population are suffering from critical protein deficiency diseases. There is, therefore, a very large potential market for animal protein foods from the sea, and the demand is bound to increase with the growth of the world's population. This offers opportunity for the development of fisheries, but, as more of the fish stocks are depleted, the need for conservation management will increase.

The pressing necessity for increasing the supply of animal protein for human consumption, which motivated the definition of conservation by the Geneva Convention, was also one reason for stressing quantity of production rather than quality, since the various kinds of fish are very nearly equal in nutritive value, although the demand for them varies widely in the market. This leads to some difficult situations where the fishermen of one nation are harvesting a given kind of fish for a high-priced market, while fishermen of another nation are supplying a large-volume, low-price market, and the latter fishery harvests incidentally a substantial share of the kind of fish pursued by the former.

A current example, which is causing international difficulty, is in the Bering Sea and recently in the Gulf of Alaska, where American and Canadian fishermen are pursuing the halibut, which has been carefully managed for a number of years to maintain the maximum sustainable yield. In recent years Russian and Japanese trawlers have been taking quantities of other species of demersal fishes, for which there is little market in the United States and Canada, but for which there is a large demand in Japan and the Soviet Union. The harvest of halibut amounts to some 75,000 tons a year, while these other fisheries are producing over a million tons a year. The large-scale trawl fisheries inevitably take a certain share of halibut, and at sizes smaller than those which produce the greatest yield per recruit. In these circumstances it may not be possible to maximize the yield of halibut and also maximize the yield of the other species which are taken in larger volume. With the increasing demand for fish protein, it would be difficult to curtail the large-volume fishery for the sake of protecting a much smaller tonnage yield from the high-quality fishery.

4. The market for many of the products of commercial fisheries is world-wide. Tuna landed in California as raw material competes with tuna shipped here originating in the Atlantic, Pacific, and Indian oceans. Buyers in California are in competition with buyers in Japan, Italy, France, and other countries. Fish meal is also an international commodity; some 2,500,000 tons of this product move in international trade.

5. The fisheries of California are not confined to the adjacent waters of the state. Our tuna fishing vessels range widely throughout the eastern Pacific from California to Chile, and into the Atlantic Ocean both off the coast of the Americas and even off the west coast of Africa. The fish processing industry in California depends for some of its raw materials, used here, on the products of a good many foreign fisheries. Companies based in California are engaged in fishing, processing, and marketing of canned tuna and other canned fish, fish meal, and, to a minor degree, in fresh and frozen fish products in a great many overseas areas, including several in Central and South America, West Africa, the Trust Territories of the Pacific, India, and

Aden, besides operations in Puerto Rico and American Samoa. We thus have a direct economic interest in the proper management of fish resources not only near our own shores but throughout the world ocean. The management of the fishery resources in our closely adjacent seas is therefore important not only with respect to the optimum utilization of those resources but also as a laboratory for developing techniques for management of sea fisheries throughout the world.

GROWTH AND POTENTIAL OF THE WORLD'S FISHERIES

The total production of the world's fisheries reached approximately 20,000,000 tons just before World War II. Production declined during the war, when naval action curtailed fishing operations, but by 1948 the world harvest had recovered to just about the prewar level. Since then, as may be seen from figure 5, reproduced from a recent publication of FAO, the growth has been very rapid; by 1962 the harvest reached 45,000,000 tons of which 40,000,000 was from the sea. The relative growth has been most rapid for two categories of fish products, the least expensive and some of the most expensive. The kinds of fish which can be caught and landed at the lowest cost are the clupeoids and the engraulids (herring, sardines, anchovies, etc.), the landings of which have more than doubled between 1956 and 1962. These species constitute the main raw materials base for the production of fish meal and oil. The very rapid growth of fish meal production is illustrated in figures 6 and 7, which show that the quantity of fish used for reduction, and the production of fish meal and solubles, approximately tripled between 1957 and 1962. At the other end of the price spectrum, the production of tuna, flatfish, and mollusks also showed substantial growth during the same period. For salmon and related species, which are highly esteemed, there was virtually no increase in production, for these resources are being fully utilized everywhere. The graphs indicate the shift taking place in the kinds of products produced for human consumption. Salted and dried cod and herring and other cured products are only holding their own or somewhat declining, but the production of frozen fish and of canned fish, exemplified by canned tuna, is increasing.

The harvest of the sea has grown from 22,000,000 tons to 40,-000,000 tons between 1952 and 1962, a rate of growth averaging 8 per cent a year in the period from 1958 to 1962. Can we expect this rate to continue? What is the potential yield of the living resources of the sea? There is no basis for answering these questions. We can only approximate the answers by examining what we know about presently unused resources, and by making theoretical calculations based on the primary productivity of the world ocean and estimates of transfer rates of organic materials between trophic levels. These approximations lead to the conclusion that the potential harvest of the kinds of organisms, including fish, mollusks, and crustaceans, that are

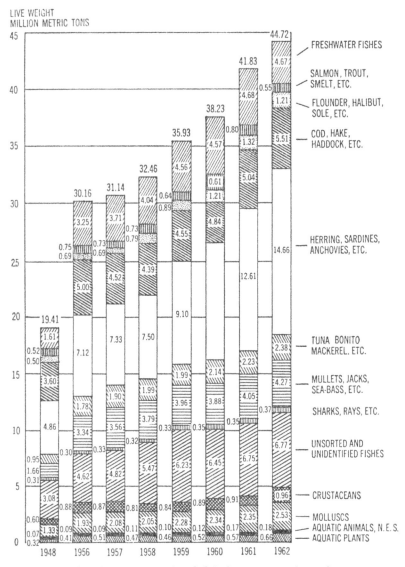

Fig. 5. World catch of fish by groups of species

now being used, and taken by foreseeable extensions of present harvesting methods, is at least greater than the present harvest by the factor of four. This does not take into account improvements which might be obtained by radical new fishing methods, such as harvesting of plankton, which are yet a long way in the future, or by radical intervention in the ecological regime, such as marine "fish farming."

On the older, fairly heavily exploited fishing grounds, a few species in greatest demand are already being harvested at a level approximat-

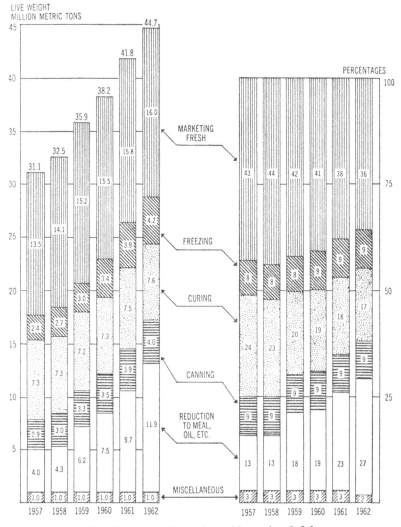

Fig. 6. Disposition of world catch of fish

ing the maximum sustainable yield, and some have been overfished. The harvesting of the Pacific and Atlantic salmon and halibut, the North Atlantic plaice and haddock, some species of tuna, some stocks of shrimp and lobsters, and the California sardine and mackerel presents serious conservation problems. However, in all these areas, with the possible exception of the nearby seas of Japan, there are populations of other fish and invertebrates which can be harvested by known methods, but are not being utilized at all or very little.

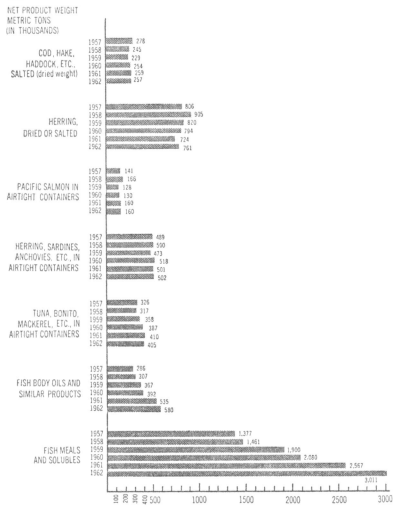

Fig. 7. Production of selected preserved and processed fishery commodities

Limiting the discussion to waters adjacent to California, we are taking only a small part of the available harvest of a large stock of jackmackerel which extends at least halfway across the Pacific. The large stock of anchovies, which has greatly increased as the sardines have declined, is almost unutilized. The very large population of hake, perhaps the most abundant harvestable species in this area, is not used, except for a few thousand pounds a year sold for mink food. The squids also are very lightly fished.

Besides the unused resources on the older fishing grounds, considerable areas of the world ocean, particularly in the tropics and in the southern hemisphere, are known to support large harvestable populations of fish and invertebrates which are being scarcely utilized. Examples are the coast of Chile south of about latitude 30°S, the Patagonian Shelf off Brazil and Argentina, and the Persian Gulf.

From all these situations it can be inferred that the sustainable production in presently fished areas could probably be doubled, and the utilization of unused resources in unexploited areas would probably double it again.

Calculations of the harvestable crop of fish and invertebrates, starting with the net primary productivity of the ocean and assuming that the harvest is, on the average, taken at the third trophic level above the phytoplankton, assuming ecological efficiency factors (transfer rates) of 10 to 20 per cent, lead to the conclusion that, at the third trophic level, there is an annual production of from 1.9×10^8 to 15.2×10^8 tons of organisms. If we assume that half of the fishery harvest is taken at the second trophic level above the phytoplankton and half at the third level, the available potentials at these two ecological efficiencies are 10.8×10^8 tons and 24.2×10^8 tons, respectively. Even though only a part of this potential can be realized, because of economic inability to harvest some of the components that are diffusely distributed, and because other predators than man take a share of the potential harvest, an ample resource base for fisheries expansion would remain. A minimum estimate of 200×10^6 tons seems reasonable and is probably conservative.

Obviously, whatever the transfer rate, the species lower in the food chain can produce a greater harvest than that of their predators. Currently, 37 per cent of the marine fishery catch consists of anchovies, sardines, herring, and the like, some of which feed almost entirely on phytoplankton and others of which feed on a mixture of phytoplankton and zooplankton; perhaps this harvest corresponds, on the average, to about one and a half steps above the phytoplankton. There appears, then, to be good reason to believe that the marine fisheries have ample room for further expansion, especially for harvesting the kinds of fish which are presently regarded as being of rather low quality.

MANAGEMENT PROBLEMS AND TECHNIQUES

In managing living resources, we may, in theory, enhance the production of the desirable resources both by modifying the physical environment and by selective removal of organisms. For living marine resources, there is little possibility, at an acceptable cost, of favorably modifying the physical environment except in the inshore waters. In enclosed bays and estuaries and in the marginal sea along the coast, such intervention is possible. Indeed, some control of the physical and biological environment has been effected for a great many years in the culture of clams, oysters, and mussels. Culture of shrimps and some species of marine fishes in ponds and embayments is practiced in various parts of the world and can be much more widely applied. Recent experiments by the California State Fisheries Department in the construction of artificial reefs to improve the habitat of certain inshore species is very promising. Conversely, the introduction of domestic and industrial wastes into embayments and the waters of the coastal zone has often adversely affected the environment; the control of waste disposal is a most important problem.

For the open sea well offshore, the very large volumes of water, the ceaseless mixing, and the migratory nature of many of the organisms generally preclude effective management of the physical environment. Only the organisms are subject to control. So far, management of high-seas fisheries has been confined almost entirely to regulating the total tonnage of the harvest of marketable species and their age or size at capture. Control of the commercial marine fisheries has been directed primarily toward maximizing the yield of a single species-population without taking into account the interactions with other harvestable populations and other kinds of organisms. Because of the importance of these interactions we need more sophisticated techniques which will consider all the important elements in the ecological system. We need to engage in a form of range management, by selective removal of organisms to maximize the combined yield of the desirable species. To some degree this may be effected by proper management of commercial fisheries. There is also the possibility of selective removal of undesirable, unmarketable species that compete with the desirable species.

The importance of considering the species interactions in the man-

agement of commercial fisheries is emphasized by the horrible example of the sardine fishery of the California Current. Because of a
complex of economic and social factors, the sardine has been selec-
tively removed while its close competitor the anchovy has been protected. The long-continuing research of the California Cooperative
Oceanic Fishery Investigations has shown that, as the sardine population declined, the anchovy population has vastly increased. At present
there is a very small population of sardines and a very small harvest,
while the population of anchovies is very large. Some reduction of the
anchovy population, which could still supply a good harvest, should
have a favorable result on the population and future harvest of the
sardine. The combined harvest of sardines and anchovies could probably be stabilized at a higher maximum level than the harvest of
either species alone.

The biological and ecological problems of managing high-seas
commercial fisheries with a view to maximizing the total production
are difficult enough. In the near-shore area, the zone from the beach
out to perhaps ten or twenty miles, the problems are vastly more difficult, because here there are several alternative and competing uses of
the living resources, as well as other uses of the ocean which may be
inimical to fish stock. We must satisfy the needs of the commercial
fisheries, and of sportsfishery and other recreation. At the same time,
the coastal zone is used for the disposal of domestic and industrial
wastes which can be harmful to the living resources. This is also the
zone where the resources of the sea floor may provide petroleum,
minerals, and other useful products which may seriously modify the
environment of the living resources. The problem, then, is to attempt
to make compatible these various uses of the marginal sea, and,
where this is not possible, to arrive at a basis of decision among the
incompatible uses. Fortunately, some of the conflicts, such as many
of those between commercial fishermen and sportsfishermen, are
more fancied than real and it should be possible to satisfy the needs
of both groups much better than we are now doing. With proper
planning for the future, we can dispose of waste in the sea and yet
maintain its living resources. Indeed, it may be possible to dispose of
waste organic material and waste heat in ways that will benefit rather
than harm the living resources. Our greatest difficulties are lack of
knowledge and lack of public understanding of the problem.

Many of the existing regulations and institutional arrangements re-
specting the uses of the living resources of the sea in the region adjacent to California are based on biological misconceptions; others flow
from conflicts of special interest groups. The closed season on taking

of abalones between the middle of January and the middle of March was established with the idea of protecting these mollusks during the supposed spawning season. But protection during the reproductive season of marine organisms which give no care to their young has little or no advantage over protection at any other time of year. Moreover, the closed season does not in fact correspond to the main reproductive season.

An example of the conflicts among different social groups is the general prohibition, except by special permit from the Fish and Game Commission, of the use of any fish except offal for reduction to fish meal and oil. This ruling was made many years ago, ostensibly on the basis that the use of fish for fish meal is somehow inferior to other uses. An attorney general of the state of California in 1927 wrote, "It seems repugnant to every right-thinking citizen to see fresh fish used for any purpose other than human consumption." I believe that this was not, and is not now, the real motive behind this ruling. Indeed, a large share of the food fish harvest in California is currently used for other than human consumption, with a goodly share of the catch of mackerel, sardines, and other food fish going into canned pet food. There is reason to believe that the real motive is the perhaps not unjustified fear that the State of California, under the present management arrangements, cannot effectively control, on a scientific basis, the harvest of the large-scale fisheries necessary for this sort of industry. The problem here would seem to be institutional rather than biological, but it is important and real.

MARINE RESOURCES PLANNING STUDY

With the rapidly increasing growth of the population and industry of California, the problems of the management of marine and other resources are bound to become more urgent and complex. There is, therefore, a pressing need to give careful consideration to the prospective growth of the state in relation to its resources, prospects, and limitations in order to provide a basis for policy decisions for orderly and optimum development. The Department of Finance, through the State Planning Office, is sponsoring a series of studies to culminate in a State Development Plan. The executive and legislative decision makers at the state level will then have a reliable, consistent, and comprehensive body of information on the trends, problems, and potentials of statewide and regional development.

As a part of the elaboration of the State Development Plan, the University of California has undertaken, through its Institute of

Marine Resources, to conduct a study of marine resources in relation to California's development. A committee of members of the faculty having specialized knowledge of marine resources will be assisted by task groups including other members of the University faculty in dealing with various categories of problems. In this ocean resources study we are undertaking the following:

1. A broad review of the role of all marine resources in California's development.

2. Evaluation of the following aspects: (a) Specific resource needs of the people of the state that could be met by the resources of the sea; (b) terrestrial alternatives to, and unique advantages of, the uses of marine resources; (c) opportunities for new uses of these resources or for expanded use of those now being utilized; (d) existing or potential conflicts in use and development of marine resources, and the nature of such conflicts.

3. Examination of the possibility of modifying the marine environment for coastal and offshore development.

4. Where appropriate and feasible, formulation of policy recommendations for the areas of public concern identified above, and indication of the information, data, and direction needed to provide the basis for further decisions.

This is the first time, so far as I know, anywhere in the world that so comprehensive and integrated a study of the multiple uses and problems of the resources of the sea has been attempted. I am hopeful that the results of this study will at least identify the problems we face in the management of marine resources, the kinds of decisions that will have to be made, and the kinds of information required as a basis of making these decisions. This pioneering venture is, I believe, a most important way in which the University can serve the people of the state, and, perhaps more important, may set a precedent for integrated study of the use and management of marine resources nationally and internationally as well.

REFERENCES

Borgstrom, George, ed.
   1962. *Fish as Food*. Vol. II. *Nutrition, Sanitation and Utilization*. New York and London: Academic Press. 777 pp.
Christy, F. T., and A. Scott
   1966. *The Common Wealth in Ocean Fisheries*. Baltimore: Johns Hopkins Press. 281 pp.

Food and Agriculture Organization of the United Nations.
    1963. *Yearbook of Fishery Statistics.* Vol. XV (1962). FAO, Rome.
Heen, E., and R. Kreuzer, eds.
    1962. *Fish in Nutrition.* First International Congress on Fish in Nutri-
        tion. London: Fishing News (Books) Ltd. 445 pp.
Institute of Marine Resources
    1965. *California and Use of the Ocean.* University of California, Insti-
        tute of Marine Resources, La Jolla, California, IMR Ref. 65-21.
Johnston, D. M.
    1965. *The International Law of Fisheries.* New Haven and London:
        Yale University Press. 554 pp.
McDougal, M. J., and W. T. Burke
    1962. *The Public Order of the Oceans.* New Haven and London: Yale
        University Press. 1226 pp.
Murphy, G. I.
    1966. "Population Biology of the Pacific Sardine *(Sardinops caerulea),*"
        California Academy of Sciences, *Proceedings,* Vol. 34, No. 1,
        pp. 1–84.
Schaefer, M. B.
    1959. "Biological and Economic Aspects of the Management of Com-
        mercial Marine Fisheries," American Fisheries Society, *Pro-
        ceedings,* Vol. 88, No. 2, pp. 100–104.
    1965. "The Potential Harvest of the Sea," *ibid.,* Vol. 94, No. 2,
        pp. 123–128.
Schaefer, M. B., and R. R. Revelle
    1959. "Marine Resources," *in* M. R. Huberty and W. L. Flock, eds.,
        National Resources. New York: McGraw-Hill Book Co., Chap.
        4, pp. 73–109.

MINERAL ASPECTS OF WATER QUALITY
IMPROVEMENT   *Everett D. Howe*

Water is the most important of the natural resources in that it is abso-
lutely essential to all plant and animal life. It is the solvent which
transports the needed chemicals through the circulatory systems to
provide both growth and maintenance of function in the organs and
cells. For men and animals it also provides the vehicle for the elimination of wastes and unneeded chemicals. Poor quality of water inter-
feres with organic processes.

While the effect of poor-quality water on plant growth is readily
observable in the vegetation in salt marshes and alkali soils, the effect
on men and animals is less obvious, since people exist in many desert
areas in spite of low-quality water. Experiments conducted during
World War II by Consolazio (1945) to determine man's tolerance
for sea water showed that the minerals not absorbed by the body are
accumulated in the kidneys and flushed out with water drawn from the
blood supply. Thus the net result of excessive mineral intake with the
sea water is dehydration of the blood stream. The United States
Pharmacopoeia standards for drinking water require not more than
500 ppm (parts per million); sheep tolerate as much as 12,500 ppm.
The extreme is reached in sea birds and a few sea animals which
utilize sea water. Some interesting observations of these birds and
mammals were published by Inoue (1963), who called attention to
the use of special glands, in the heads of birds, which perform a
desalination of sea water, and discharge the brine through ducts similar to tear ducts. However, these are very specialized cases, and animals generally require water nearly as low in minerals as does
man.

Practically all the water available for the use of man comes from
precipitation, with a very small amount possibly coming from chemi-

cal reaction deep within the earth. Water from precipitation is distilled water; the heat for evaporation from the oceans is supplied from that large nuclear reactor, the sun. Rain and snow are almost pure distilled water; the only gaseous contaminant is a relatively small amount of carbon dioxide absorbed from the air. After its contact with the earth, there is a variable amount of mineral pickup because of the solubility of many substances in water, especially in pure or distilled water. In piping distilled water in chemical laboratories, it is necessary to use lead or plastic pipes, instead of steel, to prevent corrosion. In Aruba and other locations where large distillation plants are used for water supply, minerals must be added to the distillate to "stabilize" it in order to minimize corrosive action.

It is expected that the water precipitated by natural processes will pick up some minerals in its flow along streams or in its percolation through porous formations. These run-off supplies in California pick up relatively small amounts of minerals, and reach the outlet end of aqueducts with mineral concentrations of 200 ppm or less. Even as long and winding a river as the Nile in Egypt delivers water to Alexandria, more than 1,000 miles from its source, with mineral contents of only about 250 ppm, total dissolved solids. However, in many river systems which cut deeper into the earth's surface, such as the Colorado or the large rivers of the Middle West, there is much greater increase in salinity. Other causes of increased salinity in rivers are the evaporation from surface reservoirs and the draining of waste waters into rivers. Evaporation from mountain reservoirs may amount to as much as four to six acre-feet per year per acre of water surface.

Water from wells has necessarily been in contact with soil during its percolation from the surface of the ground to the depth from which it is drawn; so it may have an appreciable mineral concentration when the soil contains soluble minerals. Examples would include the water supplies for Coalinga, California, and Webster, South Dakota. Both cities are now receiving their potable supplies from electrodialytic demineralizers. The water in the Coalinga wells has a mineral concentration of about 2200 ppm; that at Webster, about 1800 ppm.

An additional source of poor-quality drinking water comes from the mineral pick-up from many urban uses, which averages about 300 ppm. If this effluent is put back into the ground, as is the practice in some interior California cities, the underground water may increase in salinity as a result. As population grows and the demand for water increases correspondingly, man's influence on the salinity of river systems and ground waters will result in a continual decrease in the quality of the water therefrom.

DEMINERALIZING PROCESSES

Will methods for demineralizing water become sufficiently cheap to be practical for improving existing supplies of poor-quality water or for obtaining fresh water from the sea? At present the question is entirely an economic one in California, since the alternative to demineralization of local supplies is that of transporting water from distant sources. Based upon figures for the United States as a whole, Koenig (1959) has presented data showing that the main factors influencing the cost of natural supplies are the distance of transportation and the period of impoundment required to offset the variable annual precipitation on the watershed. From this information he derived curves that would indicate practical competitive costs for varying distances of water transportation. In 1959 his projection was that the small-capacity plants then conceived would only compete with the costs of water transported 100 miles or more. Since that date both the size and the cost of demineralizing plants have altered greatly, but Koenig's basic analysis approach is still valid.

While there are many possible schemes by which water can be demineralized, only two methods have become commercial: distillation and electrodialysis. On a world-wide basis about 36 mgd (millions of gallons per day) of water demineralizing capacity has been installed. Distillation in various forms accounts for about 25 mgd, and electrodialysis plants for not more than 5 mgd of the remaining. As a result of the experimental work done in recent years, the operation of modern plants is far superior to that of a decade ago. In 1952 the most economical plant produced distilled water from sea water at a cost of over $5 per Kgal (thousand gallons). In 1962 two thermal distillation plants of the Department of the Interior produced distilled water at the rate of one mgd and at a cost of about $1.40 per Kgal. The lowest cost that has been achieved for electrodialysis is at Buckeye, Arizona, where a 650,000 gpd plant is said to deliver demineralized water from a 2200 ppm source at a cost of 50 cents per Kgal (at a load factor of 48 per cent). Its cost at a load factor of 98 per cent is estimated to be 33 cents per Kgal.

Because of the low quality of many water sources in the United States, and their probable further deterioration with time, there will be greater demand for devices capable of improving water quality. Methods most appropriate for improving the quality of water from sources with initial salinities of not more than 5000 ppm include electrodialysis, ion exchange, reverse osmosis, and solvent extraction. Distillation and freeze separation are more appropriate for waters of higher salinities.

ELECTRODIALYSIS

Of the methods appropriate to the lower salinities, electrodialysis is the most advanced. It is a form of electrolysis in which the imposing of an electrical potential across a body of solution causes a migration of the ions through the solution as carriers of electricity. The equipment used for electrodialysis consists of stacks of plastic membranes spaced about 1 mm apart and sealed around the edges with gasket spacers. Water flows across the surfaces of the membranes, and a direct-current electrical potential is imposed by electrodes in contact with the two outer membranes. The plastic membranes are arranged in pairs, so that one membrane of each pair will permit only positive ions to pass through, and the other only negative ions. Water in alternate passages between the membranes is stripped of ions, while water in the intervening passages is increased in salinity. The net result is the reduction of concentration of part of the water and the increase in concentration of the balance.

A number of such plants have already been installed. The feed water to the plant at Webster, South Dakota, comes from wells and contains about 1800 ppm of dissolved minerals. The water passes through the electrodialysis stacks, consisting of 868 pairs of charged membranes, and issues from the discharge nozzle with a concentration of 275 ppm. The rate of production of the water is 250,000 gpd, and is accompanied by a flow of brine at the rate of 150,000 gpd. The water is discharged into the city mains, and the brine is sent into the sewer system. In this city, sewerage disposal is by means of a ponding system in which the liquid water evaporates under solar insolation. Thus there is an accumulation of salts in the disposal ponds, and the addition of a little more is not regarded as serious. The same amount of salts would be discharged into the sewer system with or without the treatment plant, assuming the same total quantity of water to be used. The energy required for the operation of the electrodialysis stack at Webster is only 1.7 kwhrs per Kgal as compared with about 50 to 60 kwhrs per Kgal for the most economical form of distillation. The cost of water produced at Webster is about 80 cents per Kgal.

ION EXCHANGE

Ion exchange is commonly used for the softening of water, and ordinary salt (sodium chloride) is used for regeneration. However, such usage does not reduce the total dissolved solids and hence is of no value for that purpose. The regeneration of ion exchange materials could be accomplished by the use of hydrogen and hydroxide ions; the water produced would be pure, with the exchange ions merely adding a few molecules of water. The problem here is the cost of the regenerants; the lowest cost for hydrogen ion is about $30 per ton for sulfuric acid. For an initial salinity of 1800 ppm, the amount of acid and base would be about 0.01 ton of each per Kgal and would cost something like 60 cents per Kgal of product. In an effort to avoid this expense it was suggested that a recoverable regenerant be used, such as ammonium bicarbonate, which could be separated from the product water by heating and used over again. However, investigation of this use of ammonium bicarbonate by Professor P. B. Stewart (1960) showed that the recovery would require as much heat energy as was needed for the more effective types of distillation. A search for other recoverable materials is still under way.

SOLVENT EXTRACTION

The process of solvent extraction for water purification has not been economically attractive up to date. An ideal solvent for this purpose would be one which did not absorb water, but in which the salts would be more soluble than they are in water, and would therefore extract the salts from the saline solution. However, this ideal solvent has not yet been found, but several solvents have been investigated which have a solvency for water greater than that for salts. Using these solvents, experiments have been made in which water was extracted from the saline solution, leaving salts behind.

The separation of water from the solvent solution is accomplished either by temperature changes or by fractional distillation. In the former, the absorption of water is made to occur at the temperature of maximum water solubility, and the separation of pure water from the solvent by changing the temperature to that of minimum water solubility. Separation by fractional distillation is less economical than that by temperature change. Acetone will absorb water readily and reject salts almost completely, but the acetone-water solution can be separated only by fractional distillation, which requires nearly as much heat energy as would have been needed to distill the water from the original saline solution. The search continues for a solvent that will dissolve the chemicals preferentially to the water.

REVERSE OSMOSIS

Reverse osmosis is a promising new method of desalination, although it is barely in the pilot plant stage of development. Pressure is used to force water through plastic membranes which have the osmotic property of repelling salts. This process makes use of osmosis, a natural phenomenon in which biological membranes, such as cell membranes, permit water to pass from a solution on one side of a membrane to a solution on the other side, in the direction that will tend to equalize the concentration of the solutions on the two sides of the membrane. This equalizing tendency is evaluated in terms of osmotic pressure, which is proportional to the difference in concentration of the two solutions.

An example is the circulation of fluids in trees. So long as the salinity of the water in the soil is less than that of the fluid in the roots of the tree, the osmotic pressure forces water into the root system, thus supplying pressure to pump the nourishing liquid to the top of the tree. For desalination it is necessary to reverse this process and, through exerted pressure, force water from the solution of high salinity to that of less salinity, which gives rise to the term "reverse osmosis." The encouraging discovery of synthetic membranes having this osmotic property was developed in the laboratory at the University of California at Los Angeles by Dr. S. Loeb and his co-workers (Bromley, 1965).

This method is thought to be particularly favorable for solutions of low salinity, although experiments using sea water have also given good results. The most effective membranes to date have been cast from cellulose acetate, and subjected to further specialized processes. A 4500 gpd pilot plant, completed by the UCLA group, is to be tested with brackish water.

Reverse osmosis and electrodialysis both use synthetic membranes. In the former process the property of the membrane is such that water passes through the membrane; in the latter it is the salts that pass through. According to test results from the University at Los Angeles the energy required for the reverse osmosis process is approximately the same as for electrodialysis. Its chief advantage lies in its relatively simple pumping system compared with the rather complicated electrical components of electrodialysis.

DISTILLATION

The foregoing processes seem particularly suited to the treatment of water having relatively low initial salinity. For waters of higher salinity, such as sea water with its 35,000 ppm of dissolved salts, distillation or freezing seems to be superior. Distillation, the oldest and most successful method, is used in many different forms. All make use of the fact that the dissolved salts are not volatile; thus the vapor produced by heating a saline solution will be pure water vapor and will yield pure liquid water when condensed.

The various forms of the distillation process differ in the manner in which heat energy is re-used in the evaporation process. This is important because the cost of heat energy is a major fraction of the cost of distilled water. The re-use of heat energy utilizes the fact that the boiling point of water (and of saline solutions) increases with pressure. This makes it possible to condense high-pressure water vapor in a coil submerged in saline water confined at a lower pressure, and thereby cause part of the low-pressure water to evaporate. The amount of low-pressure water evaporated is only slightly less than the amount of high-pressure vapor condensed, since the latent heats are nearly the same. The low-pressure vapor may then be condensed in a coil submerged in saline water confined at a still lower pressure, and cause some of the latter to evaporate, thus re-using the heat energy a second time. This process may be repeated several times, with the economy of the process increasing with each re-use. While the greatest number of re-uses (or effects) in a commercial plant is twelve effects (Department of Interior Demonstration Plant at Freeport, Texas), Professor LeRoy Bromley (1965) of the University of California invented and tested a rotating distillation device in which the heat energy could be re-used as many as thirty times. This device has not yet become commercial because of unsolved problems of mechanical design and high capital cost. Besides these rather straightforward multiple-effect schemes for re-using heat energy, others, such as multiple-stage flash distillation and vapor compression distillation, have been used in large plants. These differ in detail from the multiple-effect scheme, but are based on the same principles.

It is significant that all the large water production plants in the world use distillation, and that the version most favored in 1965 was multiple-stage flash distillation. An example of this type is the Department of Interior Demonstration Plant at Point Loma, California.

FREEZE SEPARATION

Freeze separation is barely out of the pilot plant stage, as the first commercial plant, a 240,000 gpd plant at Eilath, Israel, has been in operation for just one year. In the United States a 200,000 gpd pilot demonstration plant has been installed at the OSW Pilot Plant Facility at Wrightsville Beach, North Carolina.

This method makes use of the fact that the partial freezing of saline water results in the formation of small crystals of ice, but the salts remain in solution. If the ice and brine are separated, the ice can then be melted to obtain pure water. The plant at Eilath, which exemplifies the simplest freeze-separation process, consists of a freezing compartment, a washing compartment, a melting compartment, and a vapor compressor. Sea water, precooled to its freezing point, is sprayed into the freezing compartment, which is maintained at a pressure equal to the vapor pressure of water at the freezing temperature. Owing to this low pressure, part of the water will flash into vapor, thereby providing the cooling effect needed for causing some of the water to freeze into ice crystals. The rest of the water, containing all the dissolved salts from the incoming sea water, forms a brine in which the ice crystals float. This ice-brine slurry is mechanically conveyed to the washing compartment, where the brine is drained off the ice crystals and the last vestiges of brine are washed from the crystals with a small amount of fresh water. The conveyor system then moves the clean ice crystals to the melting compartment, where they are melted by the condensation of the vapor produced during the freezing process. This vapor has been compressed by the vapor compressor while passing from the freezing compartment to the melting compartment, and its condensation temperature has been made higher than the melting point of the ice crystals. The combination of the condensed water vapor and melted ice crystals constitutes the fresh water produced in this type of plant.

Because of the very large specific volume of water vapor at the freezing temperature, the compressor is extremely large and expensive. To reduce the cost, one pilot plant used a liquid desiccant to absorb the water vapor. Another pilot plant blended liquid butane with the precooled sea water before it was sprayed into the freezing compartment. The vapor formed during the spray-freezing process was then butane vapor, which is much more dense than water vapor and therefore requires a relatively small compressor to compress it for delivery to the melting compartment. None of these latter schemes have resulted in commercial applications yet.

A process related to freeze-separation utilizes hydrates of certain compounds. Propane forms solid crystalline materials when subjected to specific combinations of pressure and temperature. These hydrate crystals consist of seventeen molecules of water and only one of propane. Since the temperature of formation is much higher than that of ice formation, much less precooling of the sea water is required. Investigation of this process is being evaluated in small pilot plants.

MULTIPURPOSE PLANTS

One device for lowering the cost of natural water supplies has been the application of the multipurpose concept. Part of the cost of the reservoir and dam is provided as non-reimbursable benefits attributable to flood control, recreation, and navigation. Reimbursable costs to the water user may be further reduced if the development includes a hydroelectric plant; part of the cost of the dam can then be charged against the power plant and reimbursed from charges for the sale of electric power. Parallel features for desalination plants are difficult to arrange. So far, the only plants in which some of the costs are shared are the distiller plants in which steam is used in turbines before being sent into the distiller section. The cost of the fuel is partly offset by the charges made for electrical power and for water; the proportion assigned to each of the services is determined more or less arbitrarily. Estimates of the savings accomplished by this multiple-purpose production vary from 20 per cent to 50 per cent of the fuel cost.

The other obvious means for providing offsets for the water cost would be the production of some chemical product concurrently with the water. Little work has been done in this direction, since the predominant chemical in sea water is sodium chloride, and the quantity of this material which could be produced by a single moderate-sized desalinizing plant would be so great that it would interfere seriously with the market for this chemical. Many other materials may be obtained from sea water, and there has been some interest in seeking to combine their production with that of water. One example is the effort being made to produce a fertilizer, magnesium-ammonium-phosphate, by drawing the magnesium ion from sea water. Although a technical arrangement for producing this fertilizer has been successful, the cost of phosphoric acid used in the process has been so great that the cost of the fertilizer is greater than that of the same material produced by normal methods. Thus, no generally applicable scheme for sharing the costs of producing demineralized water has yet been developed.

NUCLEAR DESALTING PLANTS

Recent developments in nuclear reactors have directed attention to the possible merit of nuclear desalting plants. Such plants would include a nuclear reactor for producing heat energy in the form of steam, steam turbines for generating electrical power, and a distillation plant which receives its heat energy by condensing the exhaust steam from the turbine. Economies from this type of plant come from several sources, including the sharing of fuel costs between power and water and the benefit derived from very large plants. As Hammond (1962) pointed out, very substantial capital cost savings can be derived from the use of very large units producing 1000 mgd (million gallons a day) of water as compared with the 1 mgd in today's plants. Similarly, economies in the cost of nuclear reactors increase greatly as the size of the reactor is increased. While no nuclear desalting plant has yet been built, much effort is being put into development of the large distillers needed and the corresponding reactors. It seems fairly certain that within the next ten years a large plant of this type could show a water cost of as little as 25 cents per Kgal.

CONCLUSION

There are many factors that seem to be causing a degradation of many water supplies, but methods of desalination are becoming more and more effective in improving water quality. A summary of much that has been accomplished in this field has been set forth in publications by the United States Department of Interior (1963) and the University of California (1964). With our rapidly growing population, the need for such endeavor becomes increasingly important.

REFERENCES

Bromley, L. A.
    1965. "Multiple-Effect Rotating Evaporator," *Water Resources Center*, Contribution No. 100, University of California.
Consolazio, W. V.
    1945. "Drinking Water from Sea Water," *Smithsonian Report*, Publication 3817, pp. 153–163.

Hammond, R. P.
  1962. "Large Reactors May Distill Sea Water Economically," *Nucleonics,* Vol. 20, No. 12, pp. 45–49.
Inoue, T.
  1963. "Nasal Salt Gland: Independence of Salt and Water Transport," *Science,* Vol. 142, p. 1299.
Koenig, L.
  1959. "Economic Boundaries of Saline Water Conversion," *Journal of the American Water Works Association,* Vol. 51, No. 7, pp. 845–862.
Stewart, P. B.
  1960. "Sea Water Demineralization by Ammonium Salts Ion Exchange," Saline Water Conversion, American Chemical Society, *Advances in Chemistry Series,* No. 27, pp. 178–192.
U. S. Department of Interior
  1963. *Saline Water Conversion Report for 1963.* 187 pp.
University of California
  1964. "1964 Progress Report: Berkeley and San Diego Campuses," *Saline Water Conversion Research Report,* No. 65-1 (WRC No. 94). 51 pp.

PUBLIC HEALTH IMPETUS FOR AIR QUALITY
MANAGEMENT   *Bernard D. Tebbens*

For several hundred years the suspicion has existed that air pollution has a deleterious effect on health. The first Queen Elizabeth is said to have prohibited the burning of coal in London while Parliament was in session in order to eliminate the possible noxious influence of smoke and other by-products of combustion on the gentlemen assembled to legislate for the Empire. In those days there was no clear understanding of the relation between air pollution and the health of individuals. However, people living in London were then, as now, unpleasantly aware of smoke particles and sulfur dioxide, the frequent by-products of coal burning.

That inhalation of such extraneous materials could adversely affect the respiratory system is a reasonable suspicion on the part of any intelligent person. It is uncertain whether in the sixteenth century anyone distinguished between the acute and the chronic health effects of air quality deterioration. In recent times, however, say the last fifty years, such distinctions have been made. It is worth while to consider both possibilities and to note the summary of a Public Health Service document written by Dr. Heimann: "Air pollution, as it exists in some of our communities, contributes significantly as a cause or aggravating factor for the following medical conditions: acute respiratory infections, chronic bronchitis, chronic constrictive ventilatory disease, pulmonary emphysema, bronchial asthma, and lung cancer." [1]

The acute effects of deterioration in air quality came to public attention very dramatically in 1948. In the fall of that year a disastrous air-pollution episode occurred in Donora, Pennsylvania. Some twenty

[1] Harry Heimann, M. D., "Air Pollution and Respiratory Disease," *Public Health Service, Publication* No. 1257 (Washington, D. C., 1964).

people died in a few days' time, and several thousand others became ill during a prolonged attack of smog which enveloped the city of Donora and the surrounding countryside. It recalled an earlier European disaster in the Meuse Valley [2] and was forewarning of one to come in London. A popular and authoritative account of the episode was written by Berton Roueché.[3] Public notice of this event led to increased concern for air quality in the United States.

At about this same time air-pollution control in the Los Angeles basin began in earnest, not because of known health effects but because of the annoyance caused by the accumulating contaminants in the city's atmosphere, which irritated the eyes and reduced visibility. Some years later, concern in Los Angeles for a possible overwhelming episode of air pollution led to promulgation of emergency standards to control air quality.

Additional evidence of acute health deterioration from air pollution was the disastrous London episode in 1952. In that crisis, which has been repeated on a smaller scale three times since 1952, some three to four thousand people were thought to have died and many more thousands became ill. The increased morbidity and mortality were particularly evident in the health department records of the City of London. It was not so noticeable at the time as a specific disaster, although hospital records showed a large influx of patients; the magnitude of the acute effects, particularly on those known to be suffering from respiratory or cardiac ailments, became evident only later.

In retrospect, evidence of probable serious consequences of air contamination in New York City has been extracted from the records there. Greenburg and his co-workers show that the city may have suffered increased morbidity and mortality associated with deteriorated air quality at least once in the 1950's.[4] Further inspection of health department records in London indicates that similar events have probably occurred there from time to time over the past hundred years. Thus numerous episodes of severe air pollution have been associated with severe respiratory and cardiac effects, in some cases leading to death and in others to temporary incapacity.

---

[2] J. Mage and G. Batta, "Results of the Investigation into the Cause of the Deaths Which Occurred in the Meuse Valley during the Fogs of December, 1930," *Chimie et Industrie*, Vol. 27 (1932), p. 961.

[3] Berton Roueché, *Eleven Blue Men* (Boston: Little, Brown and Co., 1954). See p. 194, "The Fog."

[4] See L. Greenburg, M. B. Jacobs, B. M. Drolette, F. Field, and M. M. Braverman, "Report of an Air Pollution Incident in New York City, November, 1953," *Public Health Reports*, Vol. 77 (1962), p. 7; L. Greenburg, F. Field, J. I. Reed, and L. L. Erhardt, "Air Pollution and Morbidity in New York City," *Journal of the American Medical Association*, Vol. 182 (1962), p. 161.

Scientific study of the chronic health effects of air pollution is also of rather recent origin. Although illness resulting from prolonged contact (perhaps many years) is not spectacular, public interest in possible chronic health deterioration is stimulated by manifestations like the eye irritation which accompanies smog in Los Angeles. Since smog can irritate the tender tissues of the body, it is reasonable to wonder whether this irritation in repeated doses adds up to a long-term effect, which cannot be foreseen from the immediate symptoms. This nagging possibility has resulted in increasingly detailed research by health agencies and practicing physicians for the last fifteen years, but no clear picture has yet emerged of the total effect of air-quality deterioration on health. Opposing points of view, that air pollution is or is not a cause of chronic illness, may be based on the different characteristics of air contamination as observed in various geographic locations.

Faced with the growing concern of the people of California, the legislature in 1959 charged the State Department of Public Health with the duty of preparing standards for ambient air quality and motor vehicle exhaust, taking into account not only possible health problems but also economic and aesthetic considerations. This department has done pioneering work in developing such standards. The first group of standards, published in 1959, was supplemented in 1962. Table 1 is the complete set of standards to date (including modifications of October, 1965), omitting thirteen explanatory footnotes in the official list.

Three different levels or intensities of air contamination are of concern to the State Department of Public Health: adverse, serious, and emergency. While the latter two categories are directly related to the health of individuals as stated in the definition, the first or adverse level has several applicable criteria for any chemical compound. These have to do with eye irritation, reduction in visibility, damage to vegetation, or other nuisance or economic effects rather than with specific recognizable injury to people.

The emergency level of air pollution is defined as that level of contamination at which acute illness or death will probably occur in sensitive people. Here the Health Department concern is to avoid a repetition of a Donora or London disaster. For only one chemical entity, sulfur dioxide, is an emergency concentration described. This gas is suspected of having been a factor in the acute air pollution effects both in London and in Donora, Pennsylvania.

The serious level of air pollution is described as the concentration of a contaminant at which bodily function will be altered or which

TABLE 1
CALIFORNIA STANDARDS OF AMBIENT AIR QUALITY

Pollutant	"Adverse" level Level of sensory irritation, damage to vegetation, reduction in visibility, etc.	"Serious" level Level at which bodily function will alter or chronic disease may develop	"Emergency" level Level at which acute sickness or death of sensitive persons may occur
Photochemical Pollutants Hydrocarbons Oxidant Ozone Photochemical aerosols	"Oxidant index" 0.15 ppm for 1 hour by potassium iodide method (eye irritation, damage to vegetation, visibility reduction)	_____* Not applicable _____* Not applicable	_____* Not applicable _____* Not applicable
Nitrogen dioxide	0.25 ppm for 1 hour (coloration of atmosphere)	3 ppm for 1 hour (broncho-constriction)	_____*
Carbon monoxide	Not applicable	30 ppm for 8 hours or 120 ppm for 1 hour (interference with oxygen transport by blood)	_____*
Ethylene	0.5 ppm for 1 hour or 0.1 ppm for 8 hours (damage to vegetation)	Not applicable	Not applicable
Hydrogen sulfide	0.1 ppm for 1 hour (sensory irritation)	_____*	_____*
Particulate matter	Sufficient to reduce visibility to less than 3 miles when relative humidity is less than 70 per cent	Not applicable	Not applicable
Sulfur dioxide	1 ppm for 1 hour or 0.3 ppm for 8 hours (damage to vegetation)	5 ppm for 1 hour (broncho-constriction)	10 ppm for 1 hour (severe distress in human subjects)
Carcinogens	Not applicable	_____*	Not applicable
Hydrogen fluoride	_____*	_____*	Not applicable
Lead	Not applicable	_____*	_____*

* Explanatory footnote omitted.

may lead to chronic disease. The list shows serious levels for three gases, nitrogen dioxide, carbon monoxide, and sulfur dioxide, all of which are known to exist in the atmosphere of some cities. While sulfur dioxide is not particularly a problem in California communities, carbon monoxide is a potential danger, and its average concentration is increasing gradually in city atmospheres concurrently with the increase in automotive transportation. Carbon monoxide in the Los Angeles atmosphere has more than once exceeded the quantity and duration of the serious level according to state standards. So far as is known, no illness has been associated with such episodes in Los Angeles. The implication is clear, however, that serious illness might result from continuing increases in concentration of carbon monoxide. Oxides of nitrogen are also present in increasing amounts.

The nuisance, economic, and aesthetic effects of air pollution are fully recognized in the California standards of ambient air quality and are the most completely documented. In Table 1 the adverse category is described as the concentration at which there will be sensory irritation, reduction in visibility, damage to vegetation, or similar effects. Under this heading six items are listed as capable of producing effects not specifically related to health deterioration of individuals. One of these is oxidants which may produce eye irritation, visibility reduction because of formation of aerosols, and damage to vegetation. Another, ethylene, produces known damage to vegetation at very low concentrations. Hydrogen sulfide, having an unpleasant odor at low concentration, is considered a sensory irritant. Particulate matter, whatever its source, restricts visibility and thus affects both aesthetic and economic values. Sulfur dioxide is known to produce vegetation damage at lower concentrations than those known to produce deleterious effects on human beings; and nitrogen dioxide may impart a brown color when viewed in a long horizontal path.

Although the health of individuals and communities is the primary concern of public health agencies, nevertheless, in air-quality control as in other environmental programs, they must become involved in other than traditional health programs. The World Health Organization, at its inception nearly twenty years ago, defined health as the state of physical, mental, and social well-being rather than merely the absence of perceptible disease. Within such a broad definition public health agencies are properly concerned with the effects of the environment on aesthetics, economics, and other aspects of the total community life.

While air-quality management is not exclusively in the operating realm of health departments in the United States, a large segment of it is.[5] The principal federal agency in the field is the Public Health Service in the Department of Health, Education, and Welfare. Several other federal agencies also participate: the Weather Bureau, the Bureau of Mines, the Bureau of Standards, and segments of the Department of Agriculture.

The Clean Air Act of 1963 [6] gives great responsibility to the Public Health Service for conducting research on air-quality management and for stimulating research and control of air pollution by dispensing grants to be matched by local or state governments. From the *Directory* it can be seen that at the state level, health departments account for 44 of the total of 47 air-pollution study or control agencies. Seventeen health departments have full-time staff involvement and 27 have part-time staff involvement in air pollution. Since other state agencies account for only three full-time air-pollution control organizations, it is evident that health departments have the principal responsibility at state government level.

At local levels of government, whether city, county, or multicounty in a few instances, local government agencies other than health departments have greater participation in air-pollution control. At the local level there are 143 air-pollution control agencies in health departments as compared to 153 outside health departments. Where local air-pollution control organizations are on a full-time basis, the majority of them, 62, are in nonhealth agencies and only 17 are centered in health departments. Where the local agency has part-time concern with air-pollution control, health departments are in the majority, with 126 involved as against 91 nonhealth department agencies.

Thus it appears that at the local government level, where most of the air-pollution control is accomplished, nonhealth department concern somewhat outweighs the health department involvement. This is not unexpected considering the fact that many of the known effects of air-quality deterioration are of a nuisance and economic character. The first typical action in a control program is to prohibit open dump burning. It is evident to the public as well as to the officials that a large open fire leads to smoke, ash, and bad odors, which become a nuisance at the very least and an aesthetic detriment at the most. Contamination which can be seen or smelled is usually considered

---

[5] See Air Pollution Control Association, *1965 Directory of Governmental Air Pollution Agencies* (Pittsburgh, Pa., 1965).

[6] Further federal legislation in 1965 has placed additional responsibility for air-quality management in the Public Health Service.

most important to control in the early phases of an air-conservation program.

It is fortunate for the health of the community that many air contaminants have their earliest effects upon organisms other than man. Vegetation damage and deterioration of other factors than physical health, such as reduction of visibility, darkening of paint, or other nuisance and aesthetic effects are danger signals. Conservation of air quality in order to maintain an aesthetic and comfortable environment has often led to conservation of the environment for the benefit of man's health as well as of general community values.

The best-known and most active local air-pollution control agency, that in Los Angeles County, has no physician on its staff and is not associated with a health department. Yet the enabling legislation by which it derives its power is part of the Health and Safety Code of California. Recognizing the potential for acute health deterioration in its jurisdiction the district added emergency standards for air quality in the mid-1950's. These standards, shown in Table 2, are first, sec-

TABLE 2

ALERT STAGES FOR TOXIC AIR POLLUTANTS
(In parts per million of air)

	First alert	Second alert	Third alert
Carbon monoxide	100	200	300
Nitrogen oxides	3	5	10
Sulfur oxides	3	5	10
Ozone	0.5	1.0	1.5

ond, and third alert levels of concentration for four different gases, selected by a group of physicians, toxicologists, and other biologic experts.

The rationale of the three alert stages in Los Angeles County is that with continuous and careful monitoring of several known noxious agents the public can be warned of possible hazards from increase in the concentration of one or another of them. The first level is a forewarning of the possibility of subsequent damaging concentrations. Notification at the second alert level is based on evidence of accumulating danger to health from acute effects of one or more air contaminants. The third-level alert would mean that an emergency exists and immediate action must be taken to prevent death or serious illness. Thus over the past ten years the alert-level philosophy in Los Angeles has been that the air-pollution control agency is responsible for knowing of and warning against any health problem in the area of its jurisdiction.

In the over-all picture of air-quality management, therefore, concern for public health is the driving force for air conservation. Certainly the physical well-being of a population group is the primary interest of a public health agency. If that interest appears to be submerged in some community programs by nuisance, economic or other criteria for air pollution control, it is almost certain to emerge finally as an important determinant for continuing, abandoning or intensifying the activity.

BIOCHEMICAL ASPECTS OF THE PROBLEM OF AIR POLLUTION [1]   *J. B. Mudd*

Anyone retaining either the sense of sight or the sense of smell can easily recognize that in our urban areas, air as a resource is not as natural as it used to be. Where do the pollutants come from? What is their chemical nature? And how can we eliminate them? We have almost complete answers to the first two questions, and some progress has been made on the third. These accomplishments have been made without the need for information on the health or biochemical aspects of the problem. However, research workers in medicine and biochemistry take the pessimistic but probably realistic view that control devices on stationary and mobile sources of air pollution will not be completely effective, and, as the population increases, we may expect always to have some air pollution to contend with. Biochemical investigations are designed to discover the basis for physiological action of air pollutants, and have the practical possibility of developing preventives and cures.

Atmospheric pollutants include hydrocarbons, fluoride, oxides of nitrogen, carbon monoxide, sulfur dioxide, aldehydes, ozone, and peroxyacyl nitrates. Cities in which coal burning is prevalent suffer from sulfur dioxide pollution; often the principal source is the home fireplace. In areas of high automobile density, the exhaust provides both primary and secondary hazards. The exhaust contains unburned hydrocarbons, oxides of nitrogen, and carbon monoxide. With sufficiently high intensity of light a secondary, photochemical series of reactions can convert compounds in the automobile exhaust to peroxyacyl nitrates and ozone. These compounds are the oxidants

[1] Research on which this paper is based was supported in part by a grant (AP 71) from the Division of Air Pollution, Bureau of State Services, U. S. Public Health Service.

121

found in the polluted air of Los Angeles, San Francisco, and other cities of California.

Experimental work on mice has shown that the dose of pollutant required to kill half of any set of test animals is 12 ppm (parts per million) for ozone, 92 ppm for nitrogen dioxide, and 110 ppm for peroxyacetyl nitrate during a two-hour exposure to the gas. These concentrations are much higher than those recorded in urban areas; when the difference in weight between the experimental animals (approximately 25 gm) and man (approximately 70 kg) is taken into consideration, there appears to be little chance of human death from air pollutants in urban areas. However, lethal incidents of air pollution have occurred in London (1952), Donora, Pennsylvania (1948), and the Meuse Valley, Belgium (1930), under special circumstances. Man becomes uncomfortable in polluted air at concentrations far below lethal dosage. For cases in which the respiratory system is already taxed by disease, the extra stress may be too much. Even when the subject is healthy, air pollution can reduce working efficiency.

Assuming that a badly polluted day is one in which the level of oxidants in the atmosphere rises to 0.5 ppm and stays there for two hours, we can calculate the amount of oxidant taken into the lungs of a normal adult. During the two hours, an average adult male at rest would respire approximately 1000 liters of air, and this would contain about 1 mg of ozone. Compared with a body weight of 70 kg, the amount of toxic gas is extremely small, and yet it can be very irritating to the eyes and lungs. Since the amount of pollutant causing irritation is relatively small, the compounds of the body affected by air pollutants are probably in low concentration or in confined areas.

The situation in the plant kingdom is somewhat different. The agricultural crops severely affected by air pollution, such as spinach, lettuce, and tobacco, have relatively low weight, and a large proportion of that weight, the leaves, is available for gas exchange. Naturally occurring levels of pollutants can make these crops unsalable because of lesions characteristic of the different pollutants. The most obvious and readily assessed economic damage due to air pollutants takes place in agricultural crops. Biochemical studies of air-pollution damage, having the potential of recognizing a preventive for this damage, would provide immediate economic benefit.

Apart from differences in weight between plants on the one hand, and domestic animals and man on the other, several features affect susceptibility to air pollutants. Animals have relatively close control over body temperature, pH, and the chemical composition of body fluids, but plants undergo diurnal changes in temperature, pH of the cell sap, and content of organic acids and carbohydrates. Then changes profoundly affect the responses of plants to air pollutants. Leaf damage by peroxyacetyl nitrate is much more severe when the plants are illuminated before, during, and after exposure to the toxic gas. In contrast, leaf damage by ozone is prevented by a period of darkness before exposure. The latter effect seems to be related to depletion of the carbohydrate content of the leaves, since damage to plants kept in darkness can be induced by feeding sucrose through the leaf petioles. For peroxyacetyl nitrate, however, the interaction with light is not understood. It may be relevant that plants of ecological strains from shaded habitats are damaged by light of high intensity, specifically by impairment of the photosynthetic apparatus. Some evidence has now been obtained that peroxyacetyl nitrate damages the chloroplasts of leaves, and it is then possible that further illumination may aggravate this injury. Nonphotosynthetic tissues of plants are also affected by air pollutants, particularly by inhibition of cell expansion. The cell walls of plants are composed mainly of polysaccharide, and the inhibition of growth is evidently attributable to inhibition of the enzymes of carbohydrate metabolism.

Physiological observations on plants provide several clues at the start of biochemical investigations. Although plants and animals seem to be radically different as organisms, similarities at the subcellular and especially at the molecular level are well established, and are a natural outcome of the evolutionary process. As a consequence, biochemical investigations at the molecular level are applicable to studies of air pollution toxicity in both plants and animals.

The relatively low levels of pollutant in the atmosphere would be expected to exert their effects, at the threshold level, on subcellular compounds which are also present in small amounts. This supposition would lead us to study compounds occurring in the cell for the purpose of catalysis of metabolic reactions. The biological catalysts, enzymes, are proteinaceous, with molecular weights varying from 10,000 to values exceeding 1,000,000. These polymeric molecules have the property of accelerating chemical reactions essential for life, such as the series of reactions oxidizing carbohydrate to water and carbon dioxide, concomitantly producing energy of use to the organism; or series of synthetic reactions, such as the formation of protein, which consume energy. The action of some herbicides, fungicides, and insecticides may be attributed to specific inhibition or inactivation of enzymes. Insecticides which inhibit the enzyme acetylcholine esterase prevent transfer of nerve impulses and cause death of the insect. The herbicide 3(p-chlorophenyl)-1,1 dimethylurea specifically inhibits one of the early steps of photosynthesis in which oxygen is evolved. The synthesis and use of toxic chemicals in herbicides and insecticides is designed to combine high specificity for the target organism with low toxicity for other organisms including man. But the synthesis of air pollutants is haphazard to the point of wantonness, and we should not be surprised to find that the effects on enzymes are far from specific and are potentially hazardous to all kinds of organisms.

The action of peroxyacetyl nitrate on the enzyme isocitric dehydrogenase has been studied in detail. This enzyme catalyzes the oxidation and decarboxylation of its substrate isocitric acid, to form α-ketoglutaric acid with the concurrent reduction of a nicotinamide coenzyme.

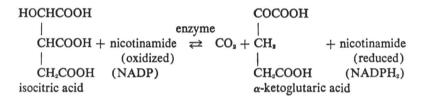

In the absence of substrate and coenzyme, the enzyme is very susceptible to inactivation by peroxyacetyl nitrate, but the presence of either substrate or coenzyme affords protection, and the presence of substrate and coenzyme together prevents damage by the toxic gas. During the process of catalysis, the substrate and the coenzyme are bound to the enzyme. So it may be concluded that the part of the enzyme affected by peroxyacetyl nitrate is protected when substrate and coenzyme are bound. The question then arises as to the chemical nature of this susceptible site. Comparison of the peroxyacetyl nitrate inhibition with inhibition by classical reagents for enzyme studies has provided circumstantial evidence that the site on the enzyme affected by peroxyacetyl nitrate is the sulfhydryl group of the amino acid cysteine.

Enzymes and other proteins are made by linkage of amino acid residues. Twenty different amino acid residues are found in proteins, and their frequency and sequence in the polymer (the primary structure) determine the helical form and configuration of the protein (secondary and tertiary structure). A segment of a protein may be represented as follows:

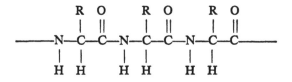

amino acid residues

When R = H—, the amino acid residue is that of glycine.
When R = CH₃—, the amino acid residue is that of alanine.
When R = CH₂SH, the amino acid residue is that of cysteine.

In isocitric dehydrogenase a functional cysteine residue seems to be concealed by substrate or coenzyme or both. These phenomena may be shown diagrammatically:

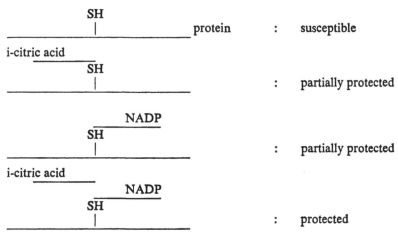

Here the isocitric acid is superior to the nicotinamide coenzyme in affording protection. In other enzymes the coenzyme is superior to the substrate.

Direct evidence has been obtained that peroxyacetyl nitrate reacts with the sulfhydryl group. (1) The enzymic activity of the proteolytic enzyme papain depends on a free sulfhydryl group. The enzyme can readily be inactivated by reaction of this sulfhydryl group with peroxyacetyl nitrate. (2) Human hemoglobin is composed of four protein chains of two different types. This tetrameric form can be dissociated first to dimers and then to monomers by lowering the pH from 7.0 to 4.5. In the tetrameric form only two sulfhydryl groups react with classical reagents such as p-mercuribenzoate. In the monomeric form a total of six sulfhydryl groups react. Therefore the conformation of the tetramer is such that four sulfhydryl groups are inaccessible to the reagent. Reaction of the sulfhydryl groups of hemoglobin with peroxyacetyl nitrate shows exactly the same dependence on degree of dissociation of the tetramer.

The reaction of peroxyacetyl nitrate with the disulfide form of the cysteine in proteins does not seem to affect enzymic activity. It is apparent from subsequent analysis of the amino acids that some reaction with the disulfide takes place, but, whatever reaction this may be, it does not change the catalytic properties of the enzyme.

The type of experimentation in biochemistry is exemplified by the reactions of peroxyacetyl nitrate. Data can be obtained under well-defined experimental conditions. But the importance of these data must then be assessed under physiological conditions which are far more complex. Is the reaction with sulfhydryl groups the basis for eye irritation and pulmonary distress in man? This information is more difficult to gather. But it is significant that reagents which react with sulfhydryl groups are eye irritants. And it has been found that the activity of certain enzymes in the lungs of animals maintained in polluted air is only 50 per cent of that in the lungs of animals maintained in clean air.

The oxidizing properties of pollutants such as ozone and peroxyacetyl nitrate are clearly the most damaging to biological material, although the ability of peroxyacetyl nitrate to cause acetylation and to form nitrite ion is another source of toxic property. Some protection of agricultural crops has been achieved by the use of ascorbic acid and other antioxidants. It remains to be seen whether further testing of antioxidants will permit their extended use to counteract air pollution.[2]

---

[2] See Seventh Annual Air Pollution Medical Research Conference, Proceedings, in *Archives of Environmental Health*, Vol. 10 (1965), pp. 141–388.

QUANTITATIVE AND QUALITATIVE VALUES IN WILDLIFE MANAGEMENT   *A. Starker Leopold*

Wild animals have many and various social values. Some people enjoy hunting as a sport, and federal and state bureaus as well as privately financed organizations are engaged in producing a shootable surplus of the common game birds and mammals. Even more people enjoy observing wildlife in its natural setting, and an equal array of governmental and conservation groups is concerned with maintaining animals for their own sake, without thought of hunting them. Some species attract tourists and hence affect local economies. Others yield commercial products such as meat, oil, furs, and ivory. Wildlife conservation in the United States has become a substantial enterprise, with complex objectives.

But the quantitative aspect of maintaining animal numbers in the face of encroaching civilization is only part of the problem. The social values attributed to wildlife often hinge upon the experience of individuals in their contact with animals, whether as hunters or as observers. And so there emerges a qualitative component, a need to preserve the atmosphere and setting in which people can enjoy their wildlife experiences (Leopold, 1954). This essay explores some of the problems and dilemmas of meeting these diverse public tastes for out-of-door adventure.

RECREATIONAL HUNTING

As the number of hunters in the United States continues to increase, the ratio of available game per hunter inevitably shrinks. An immediate and on-going problem, therefore, is to husband the existing game populations and where possible to augment them through good management.

Statistics on the increase of hunters have been assembled by the United States Fish and Wildlife Service (1961). The following abbreviated table documents the trend, both of number of hunters and the amounts of money they spend in pursuit of their sport:

	1955	1960
Number of hunters	11,784,000	14,637,000
Annual expenditure	$937,000,000	$1,161,000,000

Primary responsibility for maintaining the supply of game and regulating its use is legally vested in the state fish and game departments. The federal government plays a leading role in the management of migratory waterfowl, whose movements across state and national boundaries pose a special problem, and a secondary but important role in maintaining public domain lands (national forests, etc.) where much public hunting occurs. But the states bear the brunt of the responsibility.

State fish and game departments are funded largely or entirely by the sale of hunting and fishing licenses, whose purchase incidentally represents about 5 per cent of the sportsmen's expenditure. The system of supply, demand, and finance operates as follows. As hunters increase in number, they buy more licenses and thereby swell the departmental coffers. But they also tend to crowd the fields and forests, competing for places to hunt and game to shoot. The demand made upon a fish and game department is to supply the growing army with hunting grounds and game, utilizing the license fees to do so.

The purposes of game laws and hunting regulations are twofold: to protect the basic breeding stock of each game species, and to ration the kill of the surplus among the participating hunters. By regulating hunting in terms of local seasons, daily bag limits, and allowable methods, the kill can be kept within the limits of production; to a lesser extent the kill can be rationed among hunters. For example, by reducing the annual elk limit from two to one, the total kill will be reduced and theoretically more individual hunters will bag an animal. One of the major expenditures of fish and game departments is in the employment of wardens to enforce this phase of management. On the whole, it is perhaps the work the departments do best.

Finding a place to hunt is often more vexing to the hunter than locating game once he is afield. Departments early began to realize that they could not purchase and operate enough public shooting grounds to meet the demands of hunters. Except for limited waterfowling, the public shooting ground is impractical.

Parts of the public domain are inaccessible to the average hunter in an automobile, and considerable effort and money have been expended in the acquisition of rights-of-way and road construction to spread hunters into the back country. Whereas this represents a gain to the hunting public in one sense, it has substantially depreciated the quality of the hunting experience which formerly was enjoyed by those who were willing to reach isolated areas afoot or on horseback. The so-called "hunter access program" is a mixed blessing. Dedicated wilderness areas are theoretically immune from this development, but the Jeep and trail-bike are extending the pressures of mechanized society even to the wilderness.

The endeavor to maintain free public hunting on private lands has met with little success. In isolated rural areas, where demand is light, private lands are commonly available to the hunter. But as populations increase, especially around large urban centers, the surrounding lands tend to be closed to public access in order to protect private property. Attempts by fish and game departments to organize farm cooperatives, offering patrol and regulation of hunting as inducements, have usually failed to keep private lands open to the public for more than brief periods. On the other hand, where hunting demand is great, farmers and ranchers often begin charging admission fees or leasing their lands to hunting groups, and this promises to be a more permanent form of game utilization. When wildlife becomes a paying crop, a landowner is likely to husband it.

Suffice it to conclude here that governmental efforts to extend the hunting grounds available to the general public have had limited and qualified success.

Predatory animals live by killing other animals—sometimes game species—and it seems logical to reduce their numbers as part of a program of wildlife management. Biological research on predator-prey relations has given little support to this deduction, despite its seeming logic. In practice, predator control reduces but does not eliminate the carnivorous animals; so predation continues to affect game supplies. More importantly, however, the abundance of most prey species is limited by factors in the environment other than their natural enemies—usually by scarcity of food, cover, water, or combinations thereof. The removal of predators cannot produce more pheasants or deer if their disappearance is due to some other cause. This is widely recognized now, and most states have reduced their efforts (and expenditures) in predator control as a game protective measure. However, a few of the larger predators threaten domestic livestock and these will be subject to continued control, irrespective of game relationships. This is particularly true of the wolf, and of the coyote in sheep country.

Another aspect of the problem concerns the aesthetic value of the predators (Leopold, 1964). The presence of carnivores in the fields and woods adds something to the sport of hunting, and this value, however ephemeral, should be considered in determining predator control policy. A fleeting glimpse of a coyote or a bobcat is an extra dividend for a hunter's day.

Reducing the number of predators seemingly contributes little to game restoration and in fact may dilute the recreational value of hunting.

A direct and obvious way of increasing the game supply is to raise animals in pens and release them on the hunting grounds, thereby augmenting the natural supply. This approach to wildlife management was widely practiced by state fish and game departments in the period 1930–1955, but is no longer in favor except on private shooting preserves. Only a few game species proved amenable to artificial propagation—the ring-necked pheasant being the main one. A number of other exotic birds and some mammals were propagated with scant success. The Hungarian and chukar partridges have found niches in North America, but most exotics failed after costly efforts to introduce them. The expense of propagation is exorbitant, and survival of pen-reared birds is low. The cost of adding one pheasant to a hunter's bag often exceeds the price of his annual hunting license. Artificial propagation has been reduced in most states and completely abandoned in some, because it is an ineffectual way of meeting public demand.

But the question is rarely asked: What is the effect of game farms on the sport of pheasant hunting? It has been assumed that the success of management is measured by the sportsman's bag. But is it? Any pheasant hunter knows the difference between a wild cock and a scraggle-tailed banded bird released from a box the night before. One is a trophy, the other simply a target. Management that depreciates the quality of sport cannot be considered wholly successful. That propagation by public agencies is being abandoned for fiscal reasons is fortuitous. The matter of qualitative values has not been seriously evaluated.

Much the most effective means of assuring a game supply is to preserve, or where necessary to restore, habitat suitable to support large populations of game. This aspect of wildlife conservation is receiving more attention from government bureaus. However, it presents problems. For example, migratory waterfowl require extensive marshlands for breeding, and at least some marshlands for wintering ground. The continuing marshland drainage in the United States and Canada has steadily constricted the available waterfowl habitat. Federal, state, and private agencies concerned with waterfowl conservation are purchasing or leasing marshlands, at great cost, to preserve them for the ducks, but agricultural interests continue to drain other marshes. Preserving wetlands in an agricultural economy is neither cheap nor easily accomplished.

Wildlife production is a declared purpose of the national forests and some other areas in the public domain. Wildlife habitat is maintained by control of livestock grazing, control of wildfire, and regulated logging. But sometimes special cultural practices are needed, such as making openings in the forest canopy, restoring browse stands for deer winter range, or even controlled burning of dense chaparral. Such procedures are expensive and there is no adequate program of funding. Even on public lands, wildlife habitat is conserved only to a limited extent.

However, unlike many other management practices, habitat restoration has few adverse aesthetic effects. A well-managed forest or grazing range is a pleasant place to hunt or just to observe wildlife. To be sure, management introduces an element of artificiality in the landscape, but some degree of manipulation is tolerable except in dedicated wilderness areas, which represent a small percentage of the total recreational resources.

The precepts which have guided most aspects of wildlife management in the past are aimed primarily at the quantitative objective of supplying more hunting for more hunters, for if there were no game to hunt the sport would disappear. But in striving for this objective, public agencies have endangered the qualitative values of the sport they strive to perpetuate.

The release of hand-reared game, which changed hunting into an artificial sport, and the reduction of native faunas through predator control are no longer emphasized, not because of the adverse aesthetic effects, but because they have proved to be biologically and fiscally unsound.

Two other aspects of management may greatly affect the hunting experience of the individual: the setting of hunting seasons and the endeavor to make hunting grounds easily accessible to the public. In regulating when and where hunting takes place, game departments have it in their power to enhance or diminish sport value enormously. Some states elect to limit the deer kill by restricting the season to a few days or a week; the effect is to send into the woods an unruly mob determined to slay the deer quickly at any cost. Bullets fly, tempers flare, blood flows, and a good time is had by no one. Other states set seasons of a month to three months and regulate the total kill by other means (tags, zoning regulations, etc.). Both systems may be biologically sound, but the latter preserves the opportunity for solitude and hence retains an important component of the hunt. Similarly, developing road systems or access points may serve to concentrate hunters or to disperse them, depending on the design. The infamous "firing line" that developed at one time along the northern border of Yellowstone was a case in point. Elk that ventured over the park boundary met an array of fire power comparable to that of the Maginot line. Changed regulations and new roads completely dispersed this concentration of hunters. Hunting programs and regulations are not always set with a view to preserving the nature of sport, even though they may properly be intended to protect the supply of game.

Only the preservation or improvement of game habitat is relatively free of endangering sport values. Since it is also one of the most effective measures of preserving wildlife populations, it receives increasing emphasis today, despite costs and other limitations of land husbandry.

Despite the basic American tradition that a hunter should be able to buy a license, shoulder a gun, and go hunting, there is a marked trend toward the development of private hunting on private land, open only on a fee or lease basis. In concentric rings around major urban centers, hunting rights on private land are being sold for substantial prices.

What motivates hunters to pay dearly for a privilege that was once considered their birthright? Obviously, the sportsman who leases a duck blind or deer hunting rights on a ranch expects to improve his sport. In part, this relates to improving the probability of getting some game. But perhaps of equal importance is the element of escape from public competition, the assurance of solitude, of enjoyment of the hunt without disturbance. The hunter may be paying to recapture a qualitative value which is diminishing on public shooting grounds. A further effect of this development is to spur better management of game on private lands. The emergence of private wildlife management for profit is therefore accruing two values at no public cost: a quantitative and a qualitative one. State game departments still tend to resist the trend toward lease-hunting on the basis that it is undemocratic and against the interest of the hunting public. On the contrary, it is perhaps one of the best guarantees of the perpetuation of sport hunting in a growing community, and the departments would do well to encourage private endeavor in wildlife management.

AESTHETIC VALUES OF WILDLIFE

The whole structure of wildlife conservation and the manner of its funding through taxing hunters inevitably dictate a program that is hunter-oriented, but in recent years a new aspect of wildlife management has begun to assume great importance. Superimposed on established legal and administrative structure is the problem of satisfying the demands of the nonhunting public for the pleasure of seeing wild animals in their natural habitat. This aspect of wildlife value has developed much more rapidly than hunting values posing an interesting and complex question in resource administration.

The simplest aspect of the problem is the management of wildlife in parks dedicated to nature preservation. More involved is the question of conflicting interests on lands devoted primarily to other uses.

The Congressional Act of 1916 establishing the National Park system specifies wildlife preservation as one of the purposes of the parks. The Park Service adopted the view that their function was to protect the parks, not to manage them; and so they protected them from logging, from grazing, from hunting, from fire, and more recently from insects. For a period they even protected the grazing animals from predators. In effect, the parks were being protected not only from man but from nature. Some highly artificial situations resulted. Bears began to frequent the garbage dumps and to rob camps and beg cookies from motorists. Deer, elk, and even moose became overly abundant and damaged their own ranges, which led eventually to a drastic reduction in their numbers. Open forests, invaded by brush and conifer reproduction (following fire control), lost their carrying capacity for animals of all kinds. In 1962 the Secretary of the Interior called for a reappraisal of management policy in the National Parks (Leopold, 1963).

The new program of park management calls for a return to naturalness in park values. Local ecosystems are to be preserved or restored to the condition in which the white man first found them, so far as this is possible. The issue is primarily one of aesthetic values. A visitor seeing Yellowstone should ideally be given a glimpse of what the West was like in the time of Bridger and Carson and Fitzpatrick. Bison are best viewed grazing unhampered across the open grasslands of the Lamar Valley, not penned up in paddocks for closer viewing by the tourists. A family of black bears should be seen grubbing in a rotten log, not in a refuse heap. There should be a modest number of moose in the willows, a scattering of elk in the aspens, and an occasional coyote mousing in the meadows.

The principle of managing parks for natural values might be applied not only to units of the national park system but also to state, county, and regional parks, and even to some designated parts of municipal park lands. This type of management is not focused solely on the large animals and the dominant plants but on whole ecosystems. An obscure moss or tiny tree frog often is as deserving of preservation on park lands as the redwood or the mountain lion.

To accomplish this level of preservation or restoration calls for a high degree of skill and sophistication in ecosystem management. This, in fact, is one of the frontiers of wildlife management today.

Parks and reserves constitute less than 2 per cent of the land area of the United States. For most people the experience of outdoor life is limited to farmlands, meadows, woodlots, or hillsides that are greatly altered from their pristine condition. These hinterlands are privately owned and are used to produce crops, livestock, or timber. Land-use practices greatly affect the long-term yield of useful products, the beauty of the countryside, and, at the same time, the variety of wildlife, which adds to the interest and attractiveness of a landscape. Thus a raw, eroding gully is destructive of soil, devoid of wildlife, and at the same time ugly. Planted to trees and shrubs, the same gully is secure from erosion and supports a stock of game, songbirds, and lesser creatures. This illustrates the concept of habitat preservation referred to earlier. Many conservation bureaus and societies strive to encourage landowners to utilize their resources with these several values in mind. Since the inception of the Soil Conservation Service during the New Deal (circa 1933), substantial progress has been made in improving land-use patterns, which in the earlier era of expansion and exploitation tended to be grossly destructive.

Perhaps the greatest opportunity for aesthetic appreciation of wildlife is in or near the cities where most people live and work. The significance of wildlife management in the urban environment is suddenly attracting attention. Raymond Dasmann (1966) states the case as follows:

Today a new wave of interest in conservation is sweeping across America, bringing new challenges to all who have been professionally engaged in conservation work. In the old conservation movement we were concerned with questions of quantity of natural resources—with saving enough forest land, with producing enough wildlife, with keeping our farms yielding enough food to meet our needs. These old conservation problems have not entirely been solved, although we have made great progress. The new conservation, however, is concerned not so much with quantity as with the quality of the human environment, with particular emphasis upon the quality of the urban environment since this is where most people live. Questions of clean air, clean water, open space, and facilities for outdoor recreation for the great masses of city people are emphasized in the new conservation drive.

The White House Conference on Natural Beauty called by the President in 1965 is one manifestation of this trend. The inclusion of ecologists on county and regional planning boards is another. Recognition of public interest in "backyard wildlife" has filtered into public policy in a way that seemed inconceivable a decade or two ago.

State and fish game departments are aware of this trend but are not sure how to meet it. Perhaps the most searching examination of the problem was made by the California Department of Fish and Game (1966) in an attempt to anticipate the future recreational needs of Californians. The plan deals comprehensively with all aspects of wildlife conservation, and points out that revenue derived from sportsmen's license fees cannot meet the broad public demands for preservation of wildlife for aesthetic values. A plea is filed for general revenue support of a wildlife program that goes far beyond meeting the demands of hunters and fishermen.

SUMMARY

Wildlife management began in the United States as an endeavor to perpetuate hunting as a field sport. Most of the techniques of management were directed to increasing the supply and the yield of game species. Imperceptibly over the years the objectives have broadened to include conservation of nongame species and rare or vanishing forms whose value to society is primarily aesthetic. Game species and predators as well have acquired this new value. The recent population boom greatly accelerated the shift in emphasis from quantitative yield to qualitative appreciation. Wildlife management in the future must be realigned and refinanced to meet a very different set of objectives.

REFERENCES

California Department of Fish and Game
    1966. *California Fish and Wildlife Plan.* 4 vols. Resources Agency,
        Sacramento.
Dasmann, R. F.
    1966. "Wildlife and the New Conservation," *Wildlife Society News,*
        Vol. 105, pp. 48–49.

Leopold, A. S.
   1954. "Preserving the Qualitative Aspects of Hunting and Fishing,"
        Conservation News, Vol. 19 (20), pp. 1–5.
   1963. "Study of Wildlife Problems in the National Parks," *North
        American Wildlife Conference, Transactions*, Vol. 28, pp. 28–
        45.
   1964. "Predator and Rodent Control in the United States," *ibid.*, Vol.
        29, pp. 27–49.
U.S. Fish and Wildlife Service
   1961. "1960 National Survey of Hunting and Fishing," *USFWS Circu-
        lar*, No. 120. Washington, D.C.

Part III   DIRECTIONS FOR RESEARCH
AND POLICY

THE ECOSYSTEM AS A CONCEPTUAL TOOL IN THE MANAGEMENT OF NATURAL RESOURCES   *Arnold M. Schultz*

Ecology and economics should be very similar sciences, according to their Greek roots. They are, respectively, the study and the management of households. The etymological difference is not one of "basic" versus "applied." Indeed, the term "bionomics," supposedly a synonym of ecology, suggests that there is no important difference between *-logy* and *-nomics*.

Despite their connate derivation, today ecology and economics are not usually thought of as having anything in common. The one deals with living things in their natural environment; the other deals with production and use of resources. Neither statement is adequately definitive, but both serve to show that there is little if any overlap as the two fields are commonly conceived.

One reason for this is the equivocal role assigned to man by ecologists compared with the assignment made by economists. The term "natural" implies that there is no effective disturbance by man. Man excludes himself from nature as he chooses, or at most participates on tiptoes. When his positive relations to the environment are stressed, the specialized term "human ecology" is employed, an unnecessary distinction if man were sure of his role. In contrast, the term "resource" carries the implication of a planning agent, a manager with a goal. Here man, individually or as one of his social groupings, appraises the usefulness of his environment; in this role he makes decisions on whether or not and how to make effective disturbances in nature.

A useful concept, recently developed, reaffirms the close connections between the two *eco*-sciences better than is indicated by the original Greek roots. Appropriately, it is called the ecosystem—a functional unit conveniently organized to study in nature the activities of production, distribution, exchange, and consumption. It appears to be a conceptual framework on which not only nature can be espaliered but into which man will fit as indifferent observer, dependent participant, or independent manipulator. My main objective is to show that it is logical, both on practical and philosophical grounds, to consider the manageable unit of human, cultural, and natural resources as an ecosystem.

There are two secondary objectives. One is to straighten out the sometimes bewildering distinction between ecosystem and community that pervades the literature of ecology today. The other is to make a case for ecosystem study as a separate science. It is not botany, not zoölogy, not even ecology—it is broader than any of these; in fact, in no conceivable way is it a part of any existing scientific discipline. The implications here are rather astounding, because such a science, if that stature is granted, provides a framework for analyzing any organization integrated above the level of the individual organism. Already there are theories and methodologies which fit human societies, cultural institutions, and natural ecosystems alike. The question is whether to call this subject an interdisciplinary approach or a new discipline.

HISTORICAL BACKGROUND

Of the many terms in ecological literature, "ecosystem" [1] is one of the newest and most popular. As with many biological concepts, the term is much newer than the idea. As early as 1887 the idea was expressed as "microcosm." Tansley invented the term "ecosystem" in 1935.[2]

The ecosystem notion is by no means a creation of only the scientific intellect: in unsophisticated form it has arisen spontaneously and frequently from the thoughtful laity. This is evinced by the many frugal farmers who would rather make compost than money.

The part of Tansley's definition that is most often quoted is: "[the ecosystem is] the whole system, including not only the organism-complex, but also the whole complex of physical factors forming what we call the environment." Far more important in Tansley's conception is his statement that "ecosystems form one category of the

---

[1] "Ecosystem" is quoted when it refers to the term but not when it refers to the concept or the object; the two latter uses will be distinguished by context. "Community" is treated the same way.

[2] A. G. Tansley, "The Use and Abuse of Vegetational Concepts and Terms," *Ecology*, Vol. 16 (1935), pp. 284–307.

multitudinous physical systems of the universe, which range from the universe as a whole down to the atom. The whole method of science is to isolate systems mentally for the purposes of study, so that the series of isolates we make become the actual objects of our study."

Lindeman gave "ecosystem" a more formal definition than did Tansley: "A system composed of physical-chemical-biological processes active within a space-time unit of any magnitude."

Many scientists have made important contributions to ecosystemology.[3] These include pedologists, economists, sociologists, engineers, and chemists, in addition to ecologists. The field has developed rapidly in the last ten years, but in a normal, methodical way. This has been psychologically appealing to ecologists and other scientists and is reason enough for the ecosystem to become popular. But another factor has been far more important in making "ecosystem" the most fashionable word in the ecological literature today: just when biologists began to talk up ecosystems they began to talk down communities. "Community" has had some antagonists right from the start. Now, many "youngsters" are ready to hitch on to the "ecosystem" star. No one seems to be sure any more whether it is legal or even ethical to abstract the community, let alone do research on the abstraction.

A surprising amount of the recent literature purporting to be about ecosystems is really about communities. Some authors define "ecosystem" just as Tansley or Lindeman do, but they are using only the term and not the concept; their research and descriptions are still community-oriented.

It is not enough to clarify the ecosystem concept with a textbook definition. There needs to be a philosophy—a logical development of the notion of ecosystem which is not only conceptually independent of the community but clearly distinct from an ecologic point of view as well.

It would be difficult to develop the idea of the ecosystem without recognizing the influential role that physics has played in biological thought. The physical sciences enjoy a historic primacy; as a consequence, the physicists' method of scientific inquiry became *the* scientific method and, as another consequence, biologists were brainwashed into adopting the methods of reduction and analysis, seeking all explanation in physical terms. There are no scientific laws at the

---

[3] Ecosystemology is a highly provisional proposal. In an earlier paper I proposed "ecosystematics" and in whimsier moments have toyed with "ecosystemantics." "Biogeocoenose" offers nothing suitable. The name should be as short as "botany" and as forceful as "physics."

biological level, theories are untestable, and the phenomena too inde-
terminate for prediction and assignment of causes—these were the
arguments with which the reductionists pecked at the biologists'
efforts. Now, of course, the biologists have their own pecking order,
with the biophysicist at the "mechanistic" end and the ecologist, quite
bloody, at the other.

This reductionist tradition has impeded progress in the biological
and sociological fields. The intuitive, comfortable notions of holism
and teleology have been constantly beaten back. The conventional
line is that good scientists should ask only "How?" never "What
for?"

"What for?" is the very question for which the systems concept is
ideally designed. Life scientists have come to grips with it under the
heading of adaptation. Call it adaptation, goal-seeking, or purpose,
this question is just as intriguing with ecosystems as it is with individual organisms or, for that matter, with any level of organization.

This brings us to the theory of integrative levels, which puts
mechanism and purpose in their proper relationship. It introduces the
principle of qualitative emergence.

LAWS OF THE INTEGRATIVE LEVELS

Some of the uniformities found among the levels of organization were
set forth by Feibleman.[4] Only those most relevant to the development of the ecosystem concept will be listed and annotated here.

1. *Each level organizes the level below it plus one emergent quality.*
This means that the levels are cumulative upward. Molecules are comprised of atoms, crystals of molecules, and so on. Knowing only the
properties of the lower level, the emergent quality is unpredictable (e.g.,
hydrogen and oxygen atoms, both gases, produce a molecule of water.)
This law implies that everything has at least physical properties and this
accounts for the "reversed" position of the physical world in science and
philosophy.

2. *Complexity of the levels increases upward.* If the levels are cumulative upward, each must be more complex than the one below it. The increase in complexity is one of structure as well as emergent quality. Fifty
years ago atoms were the elementary particles; today, to the nuclear
physicist, they are complex systems with a wide span of some thirty-two
subsystems.

---

[4] J. K. Feibleman, "Theory of Integrative Levels," *British Journal of the Philosophy of Science*, Vol. 5 (1954), pp. 59–66. Feibleman actually gives twelve laws. I have combined some of the laws and corollaries.

3. *In any organization, the higher level depends on the lower.* The organism depends for its continuance on organs, tissues, cells, etc.; the ecosystem depends on organisms, soil, and water. The lower levels are more enduring—atoms last longer than molecules.

4. *For an organization at any given level, its mechanism lies at the level below and its purpose at the level above.* In an analysis of an organization, three levels must be considered: its own, the one below, and the one above. *How* an institution works is found on the levels below—its departments and offices; *what for* is found at a higher level—its social aims. Purpose is discoverable as vectors or adaptations built into organization. A watch can be taken apart by anybody, but without knowing what it is supposed to do, one cannot put it together from the separate pieces.

5. *The higher the level, the smaller the population of instances.* There are fewer molecules than atoms and fewer ecosystems than organisms. There are more electrons than anything else. The levels form a population pyramid. A corollary to this law is that the higher the level, the less it lends itself to generalizations and statistical predictability. Thus ecosystems are more distinctive than organisms and more difficult to predict as to behavior.

6. *It is impossible to reduce the higher level to the lower.* Since each level has its own characteristic structure and emergent quality, reduction is impossible without losing the quality. There is no greater reality in the parts than in the whole; they are equally real. Our scientific method has frequently led us to believe that the wholes are found at the level of common sense while the analytical parts are found only in laboratory procedures; consequently the former are illusive appearances while the latter are the realities. This, of course, is untrue.

These seem to be sensible propositions, but there needs to be more rigor in defining the reality, or exactly what is organized at the different levels. This gets to be a sticky problem, especially beyond the level of organism. In fact, it is this very problem that has kept the ecologist at the most-pecked end of the biologists' pecking order. Some supra-organism levels that have been proposed are plant communities, biotic communities, vegetation, populations, species, world fauna and flora, ecosystems. For the most part these have entirely different conceptual statuses. The taxonomic categories clearly have no place. Community is highly regarded by most ecologists as a "good" level of biological emergence. Rowe [5] gives a convincing argument that the only "reality" which deserves a place among the hierarchy of levels above the individual organism is the ecosystem. The essence of Rowe's argument follows.

[5] J. S. Rowe, "The Level of Integration Concept and Ecology," *Ecology*, Vol. 42 (1951), pp. 420–427.

ECOSYSTEM AS A PERCEPTIBLE OBJECT OF STUDY

In nature there are chunks of space-time or "events" which have both qualitative and quantitative properties. Events which have enduring modes of characterization are known as objects. The important criterion for a knowable object is volume. Volume is the basic component of perception. The spatial relationship of objects as volumes is the criterion by which consistency can be given to the level-of-integration hierarchy.

The temporal criterion is also necessary. For objects to have a high degree of "thinghood," both the spatial and the temporal modes of characterization must be sustained; that is, form and function must be constant or have rhythmic stability. This explains our interest in mature organisms, zonal soils, or climax ecosystems rather than in the developmental stages.

Now, when organized entities have strongly marked structural and functional characteristics, they are perceived as autonomous and stand out as natural objects of study. The examples used as the integrative levels (atom, molecule, crystal, cell, tissue, organ, and organism) qualify as such natural objects of study. Up to this point in the hierarchy, the levels are made consistent by applying the volumetric criterion. We can now rewrite one of the laws with the volumetric relationship of spatial inclusiveness.

7. *The object of study, at any level whatever, must contain, in the volume sense, the objects of the lower level, and must itself be a volumetric part of the levels above.* Each object on any given level constitutes the immediate environment (in the sense of impinging surroundings) of objects on the level below. Each object is a specific structural functional part of the object at the level above.

With this argument, it is obvious that the only integrative level above organism is the ecosystem, the space-time unit. It cannot be species, population, vegetation, or community. None of these are environment for individual plants nor are they any specific volumetric functional part of the ecosystem. Community is an abstracted category which brings together spatially separated parts that have no direct functional relationships. Only by way of the operational environment of the parts (the organisms) is there any functional relation between community members.

Each successive level is not made up of only one kind of lower-level organization. The object at each level is heterogeneous. Thus the ecosystem consists of the community plus whatever is spatially interwoven. The emergent qualities are the properties of the system: productivity, stability, cyclicity, diversity, trophic structure, entropy, and others.

Ecologists study pattern, abundance, vegetation changes, association between species, and countless other types of phenomena. These are logically attributes of communities. Describing and learning things about abstractions like community or vegetation is no different from describing and learning things about a species or a flora. Here, however, the objects of study which are perceptible remain as the individual organisms.

THE ECOLOGY AND PHYSIOLOGY OF ECOSYSTEMS

We have seen how an object is perceived primarily as an enduring volume. We say that we understand the object from the basic point of view of morphology. To add more comprehension we look inward. This is analysis and is achieved by reduction; that is, we look to a lower level of organization. The reductionist point of view for structure is anatomy: the comparable point of view for function is physiology. The question that physiology asks, "How does it function?" is directed to internal processes. We ordinarily think of the physiology of an organism, such as human physiology, but it can pertain to any integrative level, for example, cell physiology. By the same token we can speak of ecosystem physiology.

Next, we can add still more comprehension by looking outward, that is, from the object to the next higher level of integration. This we immediately recognize as the point of view of ecology. If we change one word of the classical definition of ecology,[6] from "organism" to "any system or integrative level," we have stated the ecological point of view of systems in general. The question that ecology asks, "What is its function?" is directed to external processes.

In the context of the level of organization hierarchy, physiological and ecological points of view overlap. The ecology of an object at one level relates it to the next higher level, while the physiology of that higher-level object relates to the level below. The relation of the organism to the environment, classical autecology, coincides with the physiology of the ecosystem of which that organism is a part. The

---

[6] Ecology is usually defined as the study of the relation of organisms or groups of organisms to their environment.

ecology of the ecosystem is its relation to larger systems and flux potentials, also, its relation to planning agents and to culture forms originating outside the system.

In this scheme there is no place for classical synecology. Synecology is a part of ecosystemology just as the community is a part—not a subsystem but a conceptual part—of the ecosystem. At this point the stage is set to explain my introductory statement concerning ecosystemology as an interdisciplinary approach versus that as a separate discipline. Present-day science is partitioned roughly according to the integrative levels. Physics deals with the atomic level, chemistry the molecular level, geology the crystal level, genetics the cellular level, and biology the organism level. You would expect the conceptual fields of study to be about the perceptible objects of study. The ecosystem is the next higher perceptible object in the hierarchy. Anatomy, morphology, physiology, and ecology are special points of view at the biological (organism) level, and occur traditionally as subdivisions of botany and zoology. This same relationship holds for the cellular level and, by stretching a point, could hold for the inorganic organizations as well. Thus we could speak of the physiology of an atom. It is relevant here that ecology and physiology are *points of view* and do not "belong" to any particular level.

Because of overlapping methodologies and the potential to cope with some common problems among the social, behavioral, and natural sciences and such applied sciences as engineering and agriculture, ecosystemology qualifies as an interdisciplinary field. By reason of its sharp characterization as a system and its clear-cut place in the hierarchy of organizations, it should enjoy a position among scientific disciplines on a par with biology, chemistry, and physics.

The study of ecosystems has its own substance and methods. Its substance consists of the lower-level objects *plus the emergent characteristics*. Some of its methods are shared with the study of systems in general. There are two entirely different ways of developing the ecosystem concept. The first stems from a general definition of "system"; the second, from the higher-level approach of human organization.

SYSTEMS IN GENERAL

One definition of system, shorn of any mathematical or philosophical precision is: "A set of objects together with relationships between the objects and between their attributes."[7] The objects need not be

---

[7] A. D. Hall and R. E. Fagen, "Definition of System," *Yearbook, Society for the Advancement of General Systems Theory*, Vol. 1 (1956), pp. 18–28.

physical; that is, the volumetric criterion is not necessary. Abstract objects such as mathematical equations or variables, theories, rules, laws, or strategies, and phenomena or events can be considered as components of systems. The attributes are the properties of these objects. Temperature or some other expression of the energy state, velocity, length, distance apart, values—these are attributes of objects. The relationships are the bonds that "integrate" for the observer the objects and their attributes. Thus any object of a possible set that cannot be tied in with any of the others is conceptually not a part of the system.

It would be impossible to say that any two objects in a set are not interrelated in some way, however trivial. Two perceptible objects always have spatial distance; components of an "idea set" always have value "distance." The degree of relatedness with which we are concerned (that is, the purpose of the observer's study) determines which objects are in the system and which are not. We set up criteria such as interest, importance, assumed effectiveness, or measurability. These four criteria are in themselves attributes of one of the objects of the system, namely the observer.

The environment of the system is defined in terms similar to those used in defining system: "For a given system, the environment is the set of all objects a change in whose attributes affect the system and also those objects whose attributes are changed by the behavior of the system." [8] This definition implies that the universe of objects of interest can be divided into two sets, system and environment, in any arbitrary way, but again dependent on the point of view or purpose of the observer. It would seem that the volumetric conception of system is much clearer in keeping the two sets (system and environment) distinct than the conception derived from the two definitions given above. However, this statement obviously applies only to ecosystem, not to systems in general, which could include various sets of abstractions.

With this interpretation of system and environment, the distinction between ecosystem, community, and population again becomes an issue. All three qualify as systems. Conceptually, ecosystem is unaffected. Community, as a set of individual organisms with definite spatial distance relationships and with assumed or measured "competitive" relationships, conforms to the definition of system. The difference between the community and the ecosystem is that in the former the observer chooses to relegate to the environment those spatially interwoven objects (soil, air, water) which in the ecosystem

---

[8] *Ibid.*

the observer uses as vehicles to carry energy and matter from one object to the next. The relationships themselves are measurable as correlation or regression coefficients, but are not observable in the sense that one can see or theorize about the exchange of a saponin molecule from one plant to another by way of a clay colloid bridge.

The population is a more exclusive set of objects than is the community. Interrelations between population members can be historical, geographical, or topological. Members of demes have the attribute of ability to interbreed. The population concept implies that relationships between members of two kinds of populations (organisms and carbon dioxide molecules) are of no interest to the observer. As soon as the population ecologist—and this holds for the synecologist too —studies habitat and niche, he is at once studying the ecosystem.

The foregoing conception of a system fits nicely Hans Jenny's state factor approach,[9] in which a certain number of variables, or state factors, define the state of the system. For example, in a physical system, when two state factors, temperature and pressure, are specified for a mol of a given gas, then most of its other properties are fixed, like density, average velocity of the molecules, or heat capacity. For an ecosystem the state factors are the following, broadly expressed:

1. Initial state factors: parent material, topography, and organism pool (flora and fauna); in other words, everything present at time zero, when the observer starts observing.
2. Flux potentials: gains from and losses to the environment, both energy and matter.
3. Time: the age of the system from time zero.

In this scheme the role of environment is clear. Environment produces no change in the system unless there is a concentration gradient or "potential" crossing the system boundary. The boundary is the point or face beyond which the observer, from his position "inside" the system, is no longer interested in what happens or can do nothing about it. Rainfall as a flux potential might be measured at the upper face of an ecosystem, just above the vegetation canopy. Whatever gradient exists or H_2O transformations occur between this upper face and the clouds is of no concern. (The meteorologist who *is* interested in these phenomena is, of course, studying a larger system.) As far as the system and its observer is concerned, the rainfall (measured as seasonal precipitation) is an independent variable; nothing can be

[9] Hans Jenny, "Derivation of State Factor Equations of Soils and Ecosystems," *Soil Science Society of America, Proceedings,* Vol. 25 (1961), pp. 385–388.

done about it. (Again, anyone manipulating rainfall by cloud seeding is manipulating a larger system and even he depends on an independent and variable source of water vapor from a far-out environment.)

The beauty of the general system approach is that it lends itself to mathematical description. The state of a system is describable uniquely by a finite set of state factors which, when quantified, yields a set of n numbers. The set of all points in the n-dimension hyperspace determines all possible states of the system. Then the functional relationships can be solved by means of a set of simultaneous differential equations.

ORGANIZATIONS IN GENERAL

Any distinction between "organization" and "system" at most is subtle. Where the objects of interest in the system are human, the former term is more frequently applied. At the risk of being far too general, it might be said that integration in systems (physical and biological level) is chiefly mechanical or transportive; and in organizations (human or rational levels), chiefly transmissive. Whatever the distinction may be, I wish to stress the deliberate and rational relationships of human beings as the organization concept and show how the ecosystem concept is identical.

An organization can be considered a set of members which, as a group, can perform functions that the members individually cannot. Each member is able to influence and be influenced by every other in the organization. Furthermore, each member acts so as to maximize the value of the results of his action to himself. Whenever a state of system exists in which the values to every member are maximum, that system has attained its highest organization.

If members act randomly, the organization is not highly organized. This is inefficient both for the members and for the organization. The inefficiency can be measured in terms of the uncertainty of each member about the actions of the others.

It can be assumed that social organizations exist because members get more value out of belonging than not belonging. This assumption can be very useful, too, in ecosystems for developing hypotheses and devising methods. Only because the observer does not have the "right" frame of reference does a species behave haphazardly in a community, or do nitrate molecules act randomly in a universe of samples. It was not uncertainty or difficulty in researching for the needed empirical facts that gave rise to our now well-publicized pesticide problems. It was the inexcusable absence of research whose rationale was based on the principle that "each member is able to influence and be influenced by every other in the organization."

The "organization" approach applies to community and population as well as to ecosystem. However, more is to be learned by studying relationships throughout the whole hierarchy of subsystems than by observing organisms (as objects or members) alone.

Members of human organizations have attributes not usually bequeathed to "lower-order" objects (e.g., problem-solving ability). This is not to say that problems needing solution in natural ecosystems do not get solved. They do, by the whole organization, not by just a few dominant, self-sufficient members.

METHODOLOGICAL ASPECTS OF ECOSYSTEM RESEARCH

In the past twenty years several lines of rigorous theoretical development have been followed intensively. These include cybernetics, information theory, game theory, decision theory, general system theory, organization theory, and open system theory. I can do no more than barely mention the applicability of these modern developments to ecosystem research and to systems research in general. When these theories are more generally known and seriously applied to management problems, we shall undoubtedly have a better understanding of complex organization. The constructs from these developments will become as commonplace in our teaching, research, and management as statistical theory is today.

Cybernetics is a theory of complex integrated chains utilizing the principle of feedback. Feedback provides mechanisms for goal-seeking either by self-controlling behavior or by amplification of divergences. There are two kinds of feedback. We may define a negative feedback mechanism as one by which part of the input-energy of a system is utilized at intervals to impose a check on the output-energy. When some of the input is used to generate or accelerate output, it is called a positive or runaway feedback mechanism. Cybernetics advances the hypothesis that such mechanisms explain purposive and adaptive behavior. Cybernetics does not give us new information, but, instead, a theoretical framework into which fundamental properties of a system fit.

For information theory, "information" is not a property but a function of a set of events, a measurable quantity like the dimension of length. Information is defined as a function of the ratio of the number of possible answers or states of a system to the number of final answers or states. It is expressed in logarithm form, and thus it is additive.

Although information and negative entropy are not identical, there is enough formal correspondence between the two concepts to allow us to make insightful analogies, despite warnings of the practical and theoretical pitfalls in such procedures.[10] It should be possible to measure entropy in a system simply by measuring the information content. The index of diversity in an ecosystem is an index of information; nonrandomness can also be expressed as information.

Game theory analyzes in a mathematical framework the rational competition between two or more anatagonists for maximum gain and minimum loss. This has many novel applications to species interactions and other ecosubsystem relationships.

Decision theory, already used in economics and human organizations, analyzes rational choices when a certain situation or state of a system is given along with its possible outcomes. This general method can be useful in pest-control studies, fisheries, and other resource management problems where interactions between populations and other parts of the ecosystem are very complex.

General system theory concerns the formulation and derivation of principles that are valid for systems in general. The theory recognizes structural similarities in many different fields. Sometimes the entities are intrinsically different and yet, as systems, they are isomorphic, and the same concepts, models, and laws apply. The unifying principle is that organization is found at all levels.

General systems research, under the aggressive leadership of von Bertalanffy, is new and stimulating. It cannot be "applied" as such to ecosystems, since it is a generalization of all systems. Serious students concerned with natural resources would do well, however, to follow the developments in the area of general systems.

Organization theory applies to any system that exhibits organized complexity. It is not different from systems theory except that the term is used mainly in social science. Human organizations are the usual center of interest, whether primary groups, institutions, or political organizations. By way of isomorphism and analogy, ecosystems and social systems can well be studied together. The methods will be the same. It should be recognized that the ecosystem is but one of many organizations integrated at a higher level than the human organism.

A system is closed if no material enters or leaves; it is open if there

---

[10] L. B. Slobodkin, "Energy in Animal Ecology," in *Advances in Ecological Research* (London: Academic Press, 1962), Vol. 1, pp. 69–101.

is import and export. The laws of thermodynamics are based on closed systems. A closed system must eventually attain a time-independent equilibrium state with maximum entropy. Diffusion phenomena and chemical reactions are examples. An open system attains not an equilibrium but a steady state with minimum entropy. Such a system can have a negative change in entropy because order is brought into the system.

In closed systems the final state is determined by the initial conditions, but in open systems the same final state may be reached from different initial states and in different ways (e.g., yield of grass in a pot can be the same with one plant or five, with several combinations of fertilizers, etc.). This is the concept of equifinality. The theory of open systems is a basis of Jenny's general state factor equation

$$l = f (L_o, P_x, t)$$

By quantifying the state factors or holding certain factors constant, the functional relationship can be solved.

APPLICATION TO RESOURCE MANAGEMENT

The application of ecosystem methodology to specific problems in resource management is likely to have outcomes different from those obtained by application of conventional ecologic or agronomic principles. One of the agronomic principles which has no connection with the ecosystem concept involves the law of limiting factors. Fertilizer companies and irrigation districts owe their success to this principle. Much of our great agricultural output can be credited to Liebig.

Simple analytical principles are used for managing many of our resources. When elm trees begin to die, their parasites are sprayed. When crop yields dwindle, fertilizers are applied. When illness strikes, medicine is administered. This symptom-cure approach is effective. It has kept our cities shady, kept us overfed, and kept our population growing. For the most part they are short-run control methods, but because of our great technological prowess we can match each problem with a cure almost as fast as the problem arises. We may be able to continue this procedure for a long time, sacrificing a qualitative value here and there, to be sure, but keeping the quantitative values intact, just as we want them.

The ecosystem concept is a useful tool for two reasons. First, it embraces a number of unique subconcepts which provide insightful models for research in the resource fields. Second, it gives a frame for questioning the efficacy of practices and policies which seem good on the surface and continue to go unchallenged.

The trophic level concept.—A food chain can be looked at in two ways. If we are organism-oriented, we think of a grass plant adsorbing nutrients, a sheep eating grass, a wolf eating the sheep, a vulture eating the wolf, a bacterium eating the vulture, and so on. If we are ecosystem-oriented, we look at it another way, as matter and energy moving along one or more of a number of pathways.

The electric circuit analogue has been used as a model to show how an ecosystem works.[11] A concentration of biomass is equivalent to voltage, the elimination of a trophic level is the same as a resistor, and energy flow is amperage. Imagine what would happen in an ecosystem if one whole trophic level were removed. The use of the technological electric circuit as a clarifying model for the common-sense object in nature is another example of the travesty of our overemphasis on the lower levels of organization and our comparative disinterest in the higher levels.

The concepts of energy and entropy.—How would you compare the efficiency of production of a prairie national park and an equivalent area of hybrid corn? The energy concept puts the annual weed, the acorn, the mule deer, and an hour of sunshine on a comparable basis. It frees us from the need to distinguish between artificial (manmade) and natural resources. It overrides the problem of qualitative differences in species. It is coldly quantitative and can be plugged into the economist mathematics as easily as a dollar. But there are some questions that the calorimeter will not answer. For example, how much brain energy is stored in a bushel of hybrid corn?

In resource management, we ought to be concerned with the loss in available energy (heat degradation) which occurs under one practice compared with that of another. This involves the concept of entropy. Here lies the key to efficient management of combustible resources: to minimize the entropy of the ecosystem. How to do it will be considered under the concept of steady state and the managerial philosophy of simulating natural processes.

Entropy is closely tied to the concepts of order, organization, and information. Here, information theory has produced for us models which are becoming very useful in experimental procedures and show great promise for systems control.

Mineral cycling.—Nutrient turnover in soils has long been inter-

---

[11] H. T. Odum, "Ecological Potential and Analogue Circuits for the Ecosystem," *American Scientist,* Vol. 48(1), (1960), pp. 1–8. See also Slobodkin, *op. cit.,* p. 82.

est to agriculturalists. Only a few have tackled such a study for the entire ecosystem.[12] If we know where certain elements accumulate in an unusable state, and how to unblock the flow, it might be a cheap way to restore soil fertility. In San Benito County, California, is an old chamise stand, over one hundred years since last burned, and possibly the oldest stand in existence. The soil is very low in phosphorus. A neighboring area, recently burned, with young vigorous chamise, has a phosphorus-rich soil. The state factors for the two areas are identical; the only difference is the occurrence of the recent fire. Actually, the ecosystems have the same amount of phosphorus: in one stand it is immobilized in the decadent chamise plants; in the other it is being circulated.

The concept of steady state.—All organizations at the biological level (cells, organisms, ecosystems) are open systems. They do not obey the second law of thermodynamics. In other words, they feed on negative entropy, they increase in order, they pass from a more probable (random) state to a less probable one, and they achieve a steady state.

In open systems the steady state is not a stable equilibrium but more like a steady motion. The steady-state system merely appears to be in equilibrium, while in fact there is a continuous interchange of matter and energy. For example, the forest floor of a "climax" forest remains unchanged, but only as to its measured quantity.

A system in a steady state is said to be stable. The entropy in unstable ecosystems tends to increase rather than to decrease. Stability would seem to be a desirable feature of an ecosystem.

The ecologist's concepts of succession and climax and their applicability to land management should be reëxamined. In community terms, succession is the replacement of one community by another. Climax condition has been exalted by range managers, foresters, and watershed managers with all kinds of superlatives. This attitude has resulted in as many pitfalls as triumphs in land management.

In succession, the change in community is just a physiognomic manifestation, a symptom of what is going on in the system. Succession is better viewed as ecosystem metabolism where the anabolic processes exceed the catabolic ones. When anabolism and catabolism are balanced, the system is in steady state (climax?). The matter of

---

[12] Ovington has recently reviewed the pertinent studies in forest ecosystems. See J. D. Ovington, "Quantitative Ecology and the Woodland Ecosystem Concept," in *Advances in Ecological Research* (London: Academic Press, 1962), Vol. 1, pp. 103–192.

vegetation change seems to be immaterial; the more important aspect is the change in entropy of the whole system.

A case in point is the chamise chaparral in the California Coast Range. Chamise is generally considered to be climax vegetation, and by conventional criteria this is reasonable. However, chamise is a notoriously poor watershed cover, and the soil under it is being degraded. In fact, the rich Yolo soils in the Cache Creek Valley are the spoils of the degradation process, and the poor Maymen soils in Lake County are what is left.

Relation between stability and diversity.—If in an ecosystem a species at a given trophic level has a wide variety of prey species and a wide variety of predators, any accidental disturbance to that species will be modulated. If such diversity is lacking, however, the results of the accident will reverberate through the whole system. Through mutations, immigration, competition, and other processes, diversity is being created constantly and this tends to increase until it is counteracted by the maximization of energy flow and transformation efficiency. Consequently, stability is reached at the time of maximum variety. An index of diversity then would be a measure of stability. So here is a clue on how to manipulate the ecosystem in order to minimize entropy. The more pathways along which energy can move and the more subsystems through which matter can cycle, the bigger the storage capacity of the system. This concept of a storage bin is very different from the one our Department of Agriculture has devised.

The concept of carrying capacity.—Range management has provided a term that should be used in much broader context: carrying capacity. The concept is a quantitative one, not qualitative. It deals with numbers at a given trophic level rather than diversity vertically through the trophic structure. Thus we may ask the question that Malthus did, in ecosystem language: At what level of human population is a steady state possible? It may just be that the question is answered more easily in this framework than in any other.

The concept of homeostasis.—Each of the concepts so far noted overlaps some of the others. It is appropriate to sum them up with the concept of homeostasis, the ensemble of regulating processes which maintains the system in a steady state. It is the organic application of cybernetics. Any stable system has homeostatic controls.

How can we use this principle in the manipulation of natural ecosystems? Feedback controls are not automatic in our managed systems; in natural systems they evolved. In man's system, the natural servomechanisms are knocked out and he is running it manually. He must learn to adjust the thermostat more finely. Nature overshoots what man wants. When he finally is aware of what has happened (for example, after the crop forecast appears) he brews a potion for counterattack. Of course, there is seldom enough time to determine the minimum effective dosage; so this time man overshoots. Nature responds valiantly, but the entire procedure can cause violent oscillations, dangerously close to being lethal.

As stable ecosystems evolve in nature, information is stored. This is a learning process. When we build our artificial systems we must put in all we know. Our research should be geared to what might happen—to anticipate the probable adaptations as well as to explain the mechanics. Then we can build self-correcting devices right into the system.[13] We can take a little solace—not much but a little—in that *Silent Spring* was written from hindsight.

The specific practices of resource management can be attacked or defended depending on which of two broad land-management philosophies is used. One philosophy is that of simulating natural processes; the second holds that modern man through his creativeness can introduce more order to ecosystems than nature can do or has done—at least he creates more order than disorder.

Simulating nature in land management.—Elsewhere I have defined agriculture as man's fattening and harvesting of a trophic level.[14] Depending on the thoroughness of the harvest, the trophic structure of the ecosystem is interrupted. All competitors at man's own trophic level are eliminated. The natural tendency of systems to increase in diversity is controlled by weeding, selection of pure strains, and so on. The ultimate in this direction is monoculture, a single species crop. Its advantages are easy to see.

In the cornfield all the available energy and nutrients being used are channeled into the corn plant. All competitors at the primary producer level (weeds) are kept out. All competitors at man's trophic level (corn borer, etc.) are kept out. Man harvests the whole crop. The highest carbohydrate production on record comes from a single crop: sugar cane.

This tremendous output does not all come from the sun. Think of the effort to keep seeds free of contamination, in selecting efficient

---

[13] Research of this kind is elegantly handled with electronic computers. With appropriate mathematical models, much exploratory work can be done inexpensively. Imaginative hypotheses and a handful of parameters are needed. See the works of K. E. F. Watt.

[14] A. M. Schultz, "The Application of the Ecosystem Concept to Range Management," California Section American Society for Range Management, Annual Meeting, Fresno, California, *Proceedings*, 1960, pp. 20–29.

strains, in building row-crop cultivators, the effort to keep biomass moving along but one pathway, the constant effort to combat nature. If this energy were subtracted from the corn crop, some of our natural ecosystems would compare favorably in productivity.

What of the nutrients and energy that cannot be utilized by the crop? Might not some other organism with different growth habits use them? Farmers used to grow pumpkins under the corn. Many plants and also animals growing together do not compete because they are drawing on different resources; they are not in the same food chain.

A few instances of complementary action—legumes and grasses, conifers and oaks—are known, but many others must exist. Even where they are suspected, the tendency in agriculture is to reduce the variety as much as possible. For example, foothill ranchers have an emotional block about brush, which is reflected in the policy of the Agricultural Conservation Program of making subsidy payments for brush clearance only when every last shrub is gone.

Besides the possibility that monoculture may not be the most efficient way to use all the influx energy, there is another drawback which stems from the relation between diversity and stability. Many serious disease and insect outbreaks can be traced directly to the absence of alternate hosts and the lack of variety in predator populations.

The fact is that we are afraid of complexity, in research as well as on the farm. That is why we build phytotrons, study ecosystems in Alaska, and teach courses that segment rather than synthesize a field.

I do not propose that forthwith we grow succotash, but we should not let the farm implement and selective herbicide industry hypnotize us into believing that the one-crop system is the most productive in the long run and over the large ecosystem.

In a forest the turnover rate of energy is slower than in a grassland or a cereal crop. A large part of the mineral and carbon materials may remain immobilized for several hundred years. Nevertheless, the forest is not without a natural consumer level. Mammals, beetles, and saprophytes give normal pathways for energy and matter. To the impatient observer, in a primitive forest these consumer organisms appear inconsequential, and because of their slow rate of consumption, primary biomass piles up. To use Odum's analogy, it is like an effective resistor in a circuit with voltage piling up in front of it. When the concentration is great enough, a short circuit is inevitable. A forest fire is a short circuit, a detour from the normal pathway.

The lumberman is a substitute consumer. He harvests certain portions of the crops. But his methods of harvest and its time distribution are such that short circuits are not prevented. In fact, the probability of fire is increased because the homeostatic controls have been removed.

A negative feedback mechanism is built into natural forests to prevent their complete destruction. In the primeval forest a small fuel accumulation results from normal forest development; then a light fire reduces the fuel to a "tolerable" level. Thus a part of the forest's energy is "fed back" to prevent a runaway fire at some later time.

This is nothing but forest adaptation. The forest has found survival in this cyclic stability. The so-called "fire subclimax" of ponderosa pine is much more stable than a many-aged, solidly structured "climax" stand with its susceptibility to a holocaustic fire.

The antithetic policies of fire exclusion and prescribed burning should be viewed in this context. The one, not entirely successful, leads to violent oscillations and lethal catastrophes. To make it work takes a tremendous amount of energy from outside the system (suppression forces). The other simulates nature. To be sure, it prevents the finely adjusted steady-state system that obtains in very old "climax" forests, but it aims for a gently vacillating state of steady motion, the limits of which can be tolerated by both forester and forest. It is economical in that most of its energy comes from the system itself.

Park management.—The energetics of a park, to my knowledge, have never been expounded. Economists have been slow in developing ways to measure recreation value, and recreationists have been slow in deciding what they want. I find the energy concept, when applied to park management, a most interesting approach.

As recently expressed in the Leopold Committee Report,[15] a reasonable goal for a park is to keep it in or build it back to its primitive condition. This implies preserving or re-creating maximum stability and diversity.

In a steady-state system, inputs and outputs are equal. A stable park would have the same characteristics. Present-day park policy is not compatible with this. Influx of energy is unimpeded. There are few curbs on the sandwiches and peanuts brought in and fed to animals; firewood, exotic plants, and other things are introduced from outside the system. In contrast, hunting, collecting, fuel gathering, lumbering, grazing, and other export activities are discouraged or not allowed. An example of a system in which influx greatly exceeds out-

[15] Leopold Committee Report, "Wildlife Management in the National Parks," *American Forests*, Vol. 69 (4) (1963), pp. 32–35, 61–63.

flux is a sewage disposal plant. Primitive America had nothing to resemble it.

Many other practices and policies in cropland and wildland management need to be reviewed with the philosophy of simulating nature in mind. They are so closely identified with conservation that nobody ever questions them. Crop rotation, brush conversion, and multiple use are three examples. It would be well to consider our resource units as systems, for this framework helps us to anticipate the consequences of actions derived from segmented thinking and piecemeal operations.

The philosophy that man creates order.—Pre-industrial man was closely dependent on his surroundings. During man's early history, settlements and civilizations sprang up only where soils were fertile. Today, especially in North American and western European countries, man is quite independent of his natural milieu. He can increase the fertility of the land where he wants to live.

Chemical elements essential to plant growth and human health were originally dispersed randomly, if not homogeneously, over the earth's surface. Nitrogen, in one form or another, occurs abundantly in all natural ecosystems. With the advent of commerce and man's penchant for living in cities, such nutrients have become more and more concentrated in certain geographical loci; these coincide with trade routes and population centers. The concentric green rings paling out with distance from Chinese cities show graphically how resources are distributed or can be created. It is a state of high improbability, a case of increase in order. (This is the same use of "order" discussed under the concept of entropy.) The Chinese city ecosystem has been fed negative entropy.

Jacks [16] cites examples of crop yields increasing in areas and times of high industrial activity. He interprets this as resulting from wealth produced in cities. Societies which are predominantly agricultural tend to deplete soil fertility while urban societies increase it. The surplus energy or wealth produced in cities increases the demand for produce and this makes it worth while for farmers to buy and use fertilizers.

Many other "problems" of man besides soil fertility illustrate the point that *not* "everything we touch turns to garbage." [17] Even if this

---

[16] G. V. Jacks, "Socio-economic Aspects of Soil Conservation and Fertility Maintenance," International Union for Conservation of Nature and Natural Resources, Seventh Technical Meeting, Athens, Greece, 1958.

[17] John Pairman Brown, *The Displaced Person's Almanac* (Boston: Beacon Press, 1962).

were so, garbage is high in caloric value and, because it is collected in dumps, it represents an order that the original ingesta never had.

Wildland is often defined as the remainder when urban and arable lands are omitted from our total land area. This is a catchall definition and connotes a highly variable and complex entity. Complexity there is because wildland has a variety of resources superimposed on each other; each resource, in turn, has a number of possible uses and may have either market or extra market value, or both. Because of these characteristics the study of wildland management should be stimulative to ecologically minded economists.

Ecologists (not economics-minded ones) have found wildlands stimulating to study for another reason. These lands are not urbanized and are not arable in the ordinary sense. Man-caused disturbances are essentially absent. The ecologist views this as a state of noncomplexity: for him wildlands are easy to study because variable Nature has put constraints on the manipulative ability of man. Such natural processes as succession and competition continue according to natural "laws"; help or hindrance by man complicates study.

In the ecosystem framework, man's presence is not taken for granted, nor is his contribution to energy loss or gain neglected in an analysis. Nonmanagement (e.g., protection at the boundary of the ecosystem) is still a kind of management. Plant communities or rare species cannot be protected without expending some energy or without creating a semipermeable barrier at the boundary of the protected ecosystem. Exchange and movement of members of other species are then impeded. One hundred per cent laissez faire is impossible, even in wilderness. Neither positive nor negative management of wildlands is "bad" unless it is considered that man is necessarily destructive, or if, in fact, he does create less order than disorder.

Whether man is able to create order in wildlands is often secondary to the question whether it is ethical for him to try. This point is nicely illustrated by a recent controversy over range management. One faction maintains that, since our natural grazing lands are not producing as much forage as they might, production should be increased by the intensive methods long used on arable lands: irrigation, fertilization, and even cultivation. The opposing faction feels that such agriculture would change the character of the rangeland intolerably, that increased productivity must be accomplished only by manipulation of the grazing pattern to allow certain forage species to gain advantage over the rest of the community. The argument holds that on a given piece of land certain substitutions and controls of plant and animal populations are permitted (within-system diddling), but manipulating the environment (state-factor control) violates some ecological ethic. Human population increases may give this problem a simple economic resolution. Or perhaps a change in the concept of what rangeland is will resolve the problem.

In the most literal sense, a man's ecosystem is his house. By artfully blending utility and beauty, an interior decorator can make this house comfortable and enjoyable for its occupants whether they are working, playing, or resting. The landscape architect does this in a larger system, and in a still larger system it is done by the regional planner. They combine the functional and aesthetic qualities of nature and culture. The techniques used by these designers are only partly a reconstruction of nature, a re-creation of what was there in the past. The other part is imaginative construction, a creation of things characteristic of the future. Regional planning is a manifestation of man's conception of organization and order. Its basic recipe is one part nature to two parts mortal architect: God, Geddes, and Frank Lloyd Wright.

It is doubtful that man will ever get an inferiority complex from being self-accused of always destroying his resources. Nevertheless, after constant admonishments, it is encouraging to know that he can create order if he wants to and if he knows what order is.

There is a parable of Rufus, who had with great effort converted a briar patch to a beautiful garden home. The parson called one day and said, "Rufus, you and the Lord have surely made a wonderful place out of this lot."

"I don't mean to be disrespectful, Parson," answered Rufus, "but you should have seen it when the Lord had it all to himself."

THE QUEST FOR QUALITY IN THE
ADMINISTRATION OF RESOURCES
Albert Lepawsky

In the analysis and administration of natural resources, there is a tendency to counterpoise the attributes of quantity to those of quality. Some resource economists have gone so far as to hypothesize that "resource quality declines continuously as quantity increases." [1] Other resource scientists and most resource conservationists actually decry this sort of mathematical syndrome in modern mass society. They argue that the overemphasis upon quantity and quantification subordinates and depreciates the quality and character of resources and resource uses and deflects from the quest for values and standards.

An inverse relationship between quantity and quality no doubt exists for certain dwindling stocks of natural resources. But for the resource process as a whole such a mathematical or logical proposition implies the additional assumption that "the resource conversion function is a continuous one," an assumption which would be widely contested. [2] Even where increase in quantity may be associated with deterioration in quality, it need not follow that these two factors are scalarly related or causally linked.

The growing concern for quality in the administration of natural resources arises in part from the new role of unconventional resources such as recreational amenities or space, where standards of measurement and criteria of evaluation lack precision. But the search for quality is also a consequence of the sheer pressure upon resources

---

[1] For a thorough critique of this theory, see Chandler Morse and Harold J. Bennett, "A Theoretical Analysis of Natural Resources Scarcity and Economic Growth under Strict Parametric Constraints," in Joseph J. Spengler, ed., *Natural Resources and Economic Growth* (Washington, D. C.: Resources for the Future, 1961).

[2] *Ibid.*, p. 42.

of human numbers and needs, to a point where existing ways of living are either threatened or actually being qualitatively transformed.

It is obvious that pressures of population and even economies of scale encourage both the demand for and the drain upon resources and that, as a consequence, the quality of goods and amenities may decline, at least for "the sensitive minority." [3] There is also contrary evidence that certain resources, and the goods and services dependent upon them, improve in quality over time. Still, the preoccupation of resource scientists and administrators with quantification threatens to result in the neglect of the less measurable and more normative aspects of their work.

Quantification in resource economics has found its most sympathetic response and widespread practice in the United States. This is not only the result of the new mechanization and managerialism inherent in the computer technology. It also has a cultural and ideological base. Blessed with an extraordinary abundance of resources but unrequited in our constant search for the better life, we have long agonized over the relations between quantity and quality. In an earlier generation, Henry Adams thought he had found an answer to this philosophic dilemma in the mathematics of energy. He gloried in the nation's inventory of resources, and regarded the enactment of the Geological Survey Act of 1879 as a high spot of public policy in his time.[4] However, the discovery of more exemplary foreign systems of resources administration and of population policy has led sensitive Americans of the Adams tradition to question the presumed relationship between quantity and quality in American life.

From a strictly technical and scientific point of view, quantity and quality do not necessarily constitute alternative, or even complementary, attributes of the resources of nature. They are differentiable for purposes of analysis or evaluation, but they become intertwined at every step in the process of making decisions and formulating policies. Modern administration and policy formulation, by their very nature, seek constantly for the quantitative expression of resource qualities. There need be no absolute conflict between the drive for quantity and the quest for quality in the management of resources.

In thus juxtaposing quantity and quality in the study of resource administration, we are not merely playing at paragrams. There is a

---

[3] E. J. Misham, *Welfare Economics* (New York: Random House, 1964), p. 96.

[4] Henry Adams, "A Law of Acceleration, 1904," *The Education of Henry Adams* (Boston: Houghton Mifflin Co., 1918), and "The Rule of Phase Applied to History," *The Degradation of the Democratic Dogma* (New York: The Macmillan Co., 1919).

revealing philological linkage between the terms "measure" and "policy." Etymologically derived from the term "meal" (food, land, agriculture), the word "measure" is defined as a "definite part of a progressive course of policy" or, more specifically, a "legislative enactment." Even "cybernetics," that most sophisticated version of the process of mensuration, is derived from a no less significant root: the Greek original, *kybernetes,* means "steersman" or "governor."

What are the potentialities of quantitative techniques as applied to resource administration and policy formulation? Policy is purposeful and in this sense qualitative by nature, and yet it must sometimes be expressed in quantitative terms. The administration of policies, as a rationalized form of human conduct, is even more subject to quantitative expression, increasingly so as a result of the contemporary computer. But since the effective administrator in modern bureaucracy is himself a formulator and enforcer of policy, he must in the course thereof constantly engage in quantifying quality and qualifying quantity.

GROSS RESOURCES AND RESOURCE AMENITIES

We already have a number of accepted quantitative measures for the quality of land, water, minerals, energy, and other natural resources. Soils, ores, and fuels have long been recognized as being extremely varied in quality, and there has been a continuing effort to analyze and describe them with greater precision. Lately it has become apparent that such free-flowing resources as water or air, or the amenities associated with them—resources and amenities which are peculiarly subject to public management—are not merely gross stocks but consist of varying qualities that often defy precise measurement. Indeed, one of the explicit criteria for public administration, as contrasted with private management, of resources is the comparative resistance to quantification, of the benefits and costs of publicly owned and managed resources.

For both water and air, public regulatory provisions can be enacted and enforced only after the specific polluting agents which man dumps into his watersheds and airsheds have been identified and their concentration measured with some precision. Nevertheless, even in the highly technical field of water or air-pollution regulation, a rapidly deteriorating ecology and environment will trigger quality controls even in the absence of refined quantitative measures of pollution. We have made more technical progress in measuring the degree of pollution in our water supplies, but now the suddenness and seriousness of the air-pollution problem are even more dramatically compelling us to take corrective measures.[5]

In the management of the United States hydrological resource, where sheer flows and gross stocks have long dominated attention, more emphasis is being placed on quality control. The recent report of the Senate Select Committee on National Water Resources concluded that the supply of the nation's water is sufficient in most regions, but that a shortage of water of good quality is impending in much of the country. Reverting to the issue of quantity for the sake of quality, this report warned that there will be insufficient water in the future to properly dilute the polluted waters or waste discharges in the nation's watersheds. In fact, the cost of maintaining water quality may soon exceed the normal cost of sustaining the necessary quantity of water itself.[6]

This blurring of the considerations of quality with those of quantity is especially apparent for amenity resources. Here the growing concern for quality has been expressed more recently in the campaign to improve urban areas, or to rationalize urban land use. To our concern about the future sufficiency of conventional resources for a high-consumption urban-industrial society, we are adding questions about the livability or quality of our environment, thus entering a new dimension of the resource problem. In short, the pressure to improve the quality of the American urban environment is essentially demographic, statistical, and quantitative in character.

The reaction against traditional policies which subordinate urban interests in the administration of resources has come mainly from political or "policy" circles rather than from the economy or the marketplace. The United States Supreme Court has outlawed certain resource-use segregation patterns and related discriminatory practices which aggravate the urban scene. The Supreme Court's reversal of the rurally biased apportionment of state legislatures is another example of this trend. A new breed of urban-minded presidents, intent upon shifting the emphasis of conservation from an agrarian to an urban base, has emerged on the "New Frontier" of American politics. They

---

[5] Orris C. Herfindahl and Allen V. Kneese, *Quality of the Environment: An Economic Approach to Some Problems in Using Land, Water, and Air* (Washington, D. C.: Resources for the Future, 1965); *Natural Resources: Air, Land, and Water,* California and the Challenge of Growth, No. 6 (Berkeley and Los Angeles: University of California Press, 1963).

[6] U. S. Senate, Select Committee on National Water Resources, *Report No. 29,* 87th Congress, 1st Session, January 30, 1961. See also Allen V. Kneese, "Water Quality Management by Regional Authorities," *Papers and Proceedings of the Regional Science Association,* Vol. 11 (1963), p. 230.

have appealed to the growing urban electorate for the support of policies "more concerned with the quality of goals than the quantity of goods." [7]

Yet much national policy making, exemplified by the newly established cabinet Department of Housing and Urban Development, is still dominated by quantitative criteria, and mathematical modeling in the urban housing market retains wide vogue.[8] Although the economists and administrators of Housing and Urban Development realize that both qualitative and quantitative priorities are necessary, there is a continued dependence upon solutions that "lie in improved techniques for calculating the costs and benefits to individual communities" in the larger metropolitan setting.[9]

The challenging relationship between quantity and quality, both in practice and in theory, is effectively posed in the recent speculation of a senior Department adviser: "Whatever problems may be associated with the accommodation of urban living, do they simply increase in proportion to scale or are they exaggerated—or possibly minimized —by scale? Or, do they change their character altogether? Do we require wholly new institutional mechanisms and policies as the scale of urbanization increases, or simply more of the same?" [10]

BENEFITS AND COSTS OF RESOURCE CONSERVATION

In the urban "beautification" movement of today and the rural conservation movement of fifty years ago, the White House has stimulated the pressures for purposes of both economic reform and social uplift. The conservation campaign of 1910 had some of its origins in the realm of ethics and aesthetics, but it emphasized economic objectives and appealed to economic interests. The movement of the 1960's, in contrast, seems to have had some economic bases, as in the campaign against poverty, but stronger still may turn out to be the search for some of the noneconomic values of American life. Considerations of market economics, however, may again prevail if the tra-

---

[7] From a commencement address by President Lyndon Johnson delivered at the University of Michigan. See *New York Times*, May 23, 1964, p. 1. The relation between urbanization and quality of life was analyzed by Harold J. Bennett and Chandler Morse, *Scarcity and Growth: The Economics of National Resource Availability* (Washington, D. C.: Resources for the Future, 1963), p. 254.

[8] See I. S. Lowry, *Model of Metropolis* (Santa Monica, California: The Rand Corporation, 1964).

[9] Benjamin Chinitz, "New York: A Metropolitan Region," *Scientific American*, September, 1965, p. 136.

[10] *Ibid.*, p. 148.

ditional forces affecting American policy and ideology are permitted to come into play.

A few resource economists have attempted to go "beyond" abundance and attack the very problems created by abundance. Some foresaw early the necessity for transforming the doctrine of Malthusian scarcity "from a problem of subsistence at the lower limit of man's survival to one concerned with the quality of life, with raising the upper limit to man's total welfare." [11] In this country there has been a retreat from Malthusianism. For the rest of this century, at least, we seem to have assured ourselves of the over-all adequacy of resources.[12] It would appear, therefore, to be on the demand rather than on the supply side that the higher standards of quality required by the consumer of resources are to be sought. The negative interest of the past in preventing the physical waste of resources is thus being supplemented with or supplanted by a demand for more positive values.

The demand side of resource economics has always involved more variables that resist exact measurement than has the supply side. Nevertheless, statistically based rationalizations and justifications are extensively employed for the support of the political reallocation of national resources and national budgets. It can be expected, therefore, that the benefit-cost formulations adopted by our waning resource-based and agrarian interests as a test of individual development projects will continue to be with us during the urban era.

The history of the trend toward benefit-cost quantification in United States policy making is a revealing one. Public administrators and political scientists, faced by the prevalence of "pork-barrel politics" in resource development projects, asked the economists, with their more sophisticated mathematics, to provide the statistical tools and the vigorous rationalizations necessary to reach viable decisions. Thus the mathematically minded economists moved in boldly with their new-found authority and soon dominated the decision-making process. With the assurance that is characteristic of their vigorous discipline, they recommended the preference for this project, the delay in that one, the eradication of another, according to their calculations of marginal investment priorities or imputed interest rates.[13]

[11] Bennett and Morse, *op. cit.*, p. 12.

[12] Hans H. Landsberg, Leonard L. Fischman, and Joseph L. Fisher, *Resources in America's Future* (Washington, D. C.: Resources for the Future, 1963).

[13] Significantly, one of the earliest government reports, dated 1947, dealt with "Qualitative Aspects of Benefit-Costs Practices." See Subcommittee on Benefits and Costs, Federal Inter-Agency River Basin Committee, *Proposed Practices for Economic Analyses of River Basin Projects* (Washington, D. C.: Government Printing Office, May, 1950), p. 1.

The contributions of these econometricists were tempered by occasional injections of "welfare" economics or by other concessions toward greater flexibility in allocating fiscal resources.

On the technical side, vast strides have been made, especially in hydrological research and water administration. Electronic computers are now found in the most unexpected places supplying not only vital data for further study but instant administrative reports applicable to either automatic or deliberative decision making and useful for various kinds of hydrological controls.[14]

After almost a decade of research and teaching, based upon rigorously constructed statistical models for water resources, the Harvard group now tends toward the conclusion that quantitative efficiency represents a restricted value system.[15] Although computational techniques have proved to be applicable to the selection and design of a single hydrologic project, they have so far been found to be relatively inapplicable to even the engineering requirements, let alone the economic and administrative design, of a hydrologic system of basin or sub-basin proportions.

The limitations of the benefit-cost method arise only partly from our inability to accurately measure noneconomic benefits. They also emerge out of the unreliability or irrelevance of measured benefits. These are often inflated or deflated by the human calculators, depending upon their own prior and personal assumptions about policy and administration. On the whole, a methodological bias has tended to exaggerate the importance of scientific precision relevant to the less important but more measurable details and to underestimate the significance of the unmeasurable complexities.

Realizing this, quantitatively minded resource economists sometimes make the most ingenious efforts to measure phenomena that are particularly recalcitrant to measurement, in order to find a firm basis for policy makers who must decide whether projects are for or against the public interest. No doubt, more acceptable forms of social ac-

---

[14] See the study by Edward J. Cleary on the work of the Ohio River Valley Water Sanitation Commission. Basic research and standard setting in water-pollution control are making vast strides. See Jack McKee, *Water Quality Criteria*, California State Water Quality Control Board, Publication No. 3A (Sacramento, 1964); Robert A. Taft Sanitary Engineering Center, Cincinnati, *National Water Quality Network: Annual Compilation of Data* (Washington, D. C.: Government Printing Office, 1957/58–1962/63).

[15] Arthur Maas and Maynard M. Hufschmidt, "Report on the Harvard Program of Research in Water Resources Development," in *Resources Development Frontiers for Research*, Western Resources Conference, 1959, pp. 133–179.

counting of this type will materialize in the future.[16] Meanwhile, resource policies continue to incorporate the best that economic benefit-cost computation has to offer. The Land and Water Conservation Act of 1965, for example, treats the mathematics of resources rather cavalierly, professing to calculate and consider—on the same political scale as it were—"direct and indirect costs to the Government," "benefits to the recipient," and also "public policy or interests served." [17]

The prevailing formula for making resource policy seems to be to measure by any means possible, but, by all means, measures must be passed.

BROADENING THE DIMENSIONS OF THE RESOURCE PROCESS

The greatest opportunities for the mathematical approach to resource policy formulation would seem to reside in the realm of national strategy and international administration. Extensive scientific applications in space and strategic policy are already under way. For military decision making, particularly, and in the international allocation of resources, it may be too early to predict the outcome. Nevertheless, because of the nature of the human and historical factors, over which the American mind is likely to permit sentiment as well as technology to prevail, we may expect a more balanced resource-use program in strategy and space eventually—a program which will probably be more international in scope.

Certainly, in the realm of international aid or its emerging counterpart, international trade, the United States will continue to commit both natural and fiscal resources on an enormous scale. This will no doubt invite the continuation of broad-scale experiments with quantified decision making on a world-wide scale. But there is also likely to be a parallel rise in uncomputerized policy making involving selectivity and quality.

International economics is firmly attached to per capita income statistics. But, although this capsuled mathematical formulation may be useful in indicating to what extent, globally speaking, the poor are getting poorer and the rich are getting richer, its application to the process of arriving at workable decisions and policies for national de-

---

[16] Raymond W. Mack and Dennis C. McElrath, "The Study of Man: A Computerized View of the World," *Trans-Action*, January–February, 1959, p. 27; Bertram M. Gross, "Planning: Let's Not Leave It to the Economist," *Challenge*, December, 1965, pp. 30–33.

[17] "Land and Water Conservation Fund Act of 1965," 78 Stat. (1964), 897.

velopment programs and international assistance projects is seriously limited.[18] Not only are currency conversion rates untrustworthy for such purposes, but the nonmarketable and nonmeasurable factors in underdeveloped countries are of considerable importance. Still, it is surprising how, in developing economies, scarce skills and techniques often get diverted into premature computerization when well-chosen pencil-and-paper calculations could do the job just as well.

It may be argued that deprivation is so widespread throughout the world that considerations of quantity will for some time continue to outweigh considerations of quality. Nevertheless, quality factors will continue to interweave with quantity factors. For example, the experience of the United States in converting its surplus wheat into Bulgar wheat as "ala," which looks or tastes like rice, suggests the need for more indigenous quality specifications and standards for even the most common consumer goods.[19] Certainly, local vagaries of taste are not easily internationalized. Meanwhile, the abundant societies can help by offering a wider range of qualitative choices for the varying palates of the world.

Paradoxically, in a world that suffers quantitatively from overpopulation, a major problem we shall have to struggle with is the quality of the world's human resources. In dealing with raw resources or their fabricated products, we can if we wish stick to quantitative terms. But human beings are living organisms whose values cannot be constantly quantified with impunity. If values are ignored where the human dimension is concerned, they will force their way sooner or later into the decision-making process, intruding themselves on both the quantified data and the quantifiers' objectives.

Historically, it may be well to recall that Malthusianism had a distinctly normative origin despite its precocious indulgence in the inductive method. Certainly, the secondary wave of Malthusianism, despite its hedonistic calculus, proved to be even more sensitive to values and less impressed with numbers. "A population may be too crowded," warned John Stuart Mill, "though all be amply supplied with food and raiment." [20] Comparative demography and inter-

---

[18] See esp. Harry Stark, *The Emerging World Economy* (Dubuque, Iowa: W. C. Brown Co., 1963).

[19] Research on the transformation of wheat to simulate rice is being conducted at Western Regional Research Laboratory, U. S. Department of Agriculture, Albany, California. Reports indicate an increased use of this product, and research is under way to develop a type of wheat without the qualities to which rice-eating people object, e.g. dark color, stronger taste, etc.

[20] John Stuart Mill, "On the Stationary State," *Principles of Political Economy* (London: Longmans, 1920), p. 750.

national economics today might well be as sensitive to the need for linking considerations of quantity and quality.

Are the global supplies of natural resources sufficient to provide the kind and quality of material and energies that will sustain a rewarding life and preserve the human species? Despite the promise of atomics and protonics, a Neo-Malthusianism is posting warnings about the exponential amounts of energy required in the emerging era. Although a calorie is a calorie and a kilowatt is a kilowatt, the question of energy resources should be raised in the most broadly qualitative terms possible.[21]

Ultimately, we face much more than the problem of providing sheer energy supplies at economic rates. Prudent and rational human administration using mental capacities of the first order will be equally essential. Public administration will offer a great challenge to the most rigorous scientific minds and inventive intellects that the forthcoming era is likely to produce. It may be trite but it is true to say that what is at stake here is the conservation of the most crucial of our resources: creative intelligence and the human species itself.

Other basic questions are being raised about the ultimate and overall stock of natural resources beyond any presently presumed limits of the energy resource. We are now being warned that even if world population controls are postulated, we may run out of space. And this does not refer to living space alone. The penetration of the hydrosphere and the stratosphere will add further dimensions to our thinking. But the ultimate shortage in this realm may be the egregious demands upon the finite band of wave lengths available to mankind— his communication spectrum and therefore his vital intelligence system on this planet and beyond.[22]

And so the analysis moves from the potential limits of land to those of water, minerals, energy, air, atmospheric space, and wave lengths. Steadily, ecologists and conservationists adopt a more measured pessimism,[23] but also a more cautious optimism, about the

[21] Frederick Soddy, *Matter and Energy* (London: Williams and Norgate, 1912); Fred Cottrell, *Energy and Society: The Relation between Energy, Social Change, and Economic Development* (New York: McGraw-Hill Book Co., 1955).

[22] Richard L. Meier, "Information Resource Use and Economic Growth," *op. cit.* in note 1, above, pp. 98–125.

[23] The general trend in the literature is exemplified by Fairfield Osborn, *Our Plundered Planet* (Boston: Little, Brown and Co., 1948), and *The Limits of the Earth* (Boston: Little, Brown and Co., 1953); Harrison S. Brown, *The Challenge of Man's Future: An Inquiry Concerning the Condition of Man during the Years That Lie Ahead* (New York: Viking Press, 1954), and *Next Hundred Years: Man's Natural and Technological Resources* (New York: Vik-

issue of gross available supplies, including now the potential inputs from inner and outer space.

Changing as fast as the technological variables are the political factors, with numerous and complex impacts upon the international economy. One of the most controversial issues before the United Nations next to, and associated with, the search for peace is the long-range contest between the so-called raw-materials countries and the industrial countries. There are probably few United Nations bodies whose deliberations are more explosive than those of the Commission on Permanent Sovereignty over National Resources.[24] In many respects, the continuing battle over the Congo has been a struggle for some measure of control over the administration of the resources of this naturally well-endowed land.

Issues of high significance are emerging for the proper management of world resources that combine the problems of the underdeveloped with those of the developed regions. Yet these are issues which must be handled by far less complicated methods than computerized calculation. Indeed, these problems might well become obscured by overly sophisticated methods of quantification.

PROSPECTS FOR QUANTITATIVE AND QUALITATIVE APPROACHES

So far as quantitative techniques can serve the requirements of a rational and responsible policy for natural resources, their retention can be expected. However, it would be more realistic to anticipate that the simpler kinds of calculation will prove to be more applicable for both policy and administration.

Often, the major choice is one of either yes-or-no or in-what-order. Increasingly, economists point out that the question of exactly how much or what ratio is answered by means of assumptions which are subject to controversy and challenge, and which resort to benefit-cost devices as rationalizations.[25] The net effect of such mathematical decision making and policy formulating is often merely to intensify,

---

ing Press, 1957); William Vogt, *Road to Survival* (New York: W. Sloane Associates, 1948), and *People! Challenge to Survival* (New York: W. Sloane Associates, 1960); L. Dudley Stamp, *Our Undeveloped World* (London: Faber and Faber, 1953), and *Our Developing World* (London: Faber and Faber, 1963).

[24] See reports of U. N. Commission on Permanent Sovereignty over National Resources.

[25] S. V. Ciriacy-Wantrup, "Benefit-Cost Analysis and Public Resource Development," in Stephen C. Smith and Emery N. Castle, eds., *Economics and Public Policy in Water Resource Development* (Ames, Iowa: Iowa State Uni-

without refining, the technical conflict between market and welfare economics in resource development.

One of the most broadly based resource policies of the United States—that of the continuing but sometimes "uneconomic" development of the West—would hardly have been changed had there been a precocious nineteenth-century benefit-cost economics or quantified decision making available at that time in this country.[26] In this historical sense, resource policy making will probably continue to resist the dictates of the computer. Research will tend to be concentrated not merely on more precise measurement but also upon more viable systems of decision making and more forceful devices of public administration. This calls for more inventiveness with regard to our political institutions and administrative instruments rather than a greater refinement of our mathematical methodology.

Multiple-use planning and programing of the forest resource, for timber, water, grazing, power, recreation, and wildlife purposes, demand not so much elaborate benefit-cost quantification as simple but careful scheduling of specific priorities based upon values derived from existing legislation or established policies. The use of the forest for amenity purposes may even lead to nonuse of a resource in certain areas. This is, in effect, true of the wildlife or wilderness resource, where the objective is to preserve the sheer sustaining quality of untouched space. Wilderness benefits and costs may be measurable, but they are not always economically so. Their values may be more readily related to mental repose than to market economics.

Put in its proper place, however, resource quantification can have a lasting residue of influence on the science and study of natural resources. The main adjustments needed may be more in training methods than in research. The intellectual skill required for the computerized administration of resources is quite different from the attitude of mind essential for determining sound conservation policies. Both quantitative analysis and quality-focused ideas can be fitted into a dynamic curriculum, and, at least so far as students are concerned, need not be antithetical. There is more likelihood that teaching and research will require adjustment and a more flexible in-

---

versity, 1964); John Krutilla, "Welfare Aspects of Benefit-Cost Analysis," *ibid.*, p. 23; see also the papers and subsequent discussions in Robert Dorfman, *Measuring Benefits of Government Investment: Papers Presented at a Conference of Experts, November 1963* (Washington, D. C.: The Brookings Institution, 1965).

[26] Irving K. Fox, *Major Current Policy Issues in Water Resource Development and Use*, A Report for the Western Resources Conference (Fort Collins, Colorado: The Conference, 1965).

tellectual commitment on the part of the professors. The nonmathematical minds among them may have to learn more about computation, and the mathematically inclined more about alternative methodologies and differing value systems.

This is not to imply that quantitative matters about resources should be left to the mathematical economists, and the value-laden questions should be answered by the political scientists. On the contrary, the lessons learned by the economists from their research into the mathematics of the market and by the political scientists in their application of statistics to electoral behavior should help to bring them together on resource policy.

The computer can add constructively to the grander interdiscipline of political economy. Adam Smith, who himself deferred to the transcendent "art of the legislator," is not celebrated for his mathematical digressions into the subject of agricultural marketing. Nor is Alfred Marshall remembered for his mathematical translations of Ricardo's diminishing returns. As for John Maynard Keynes, though he was originally a devotee of Malthusian demography, his main contributions, those involving public investment and full employment, were independent of his econometrics, which few have read and fewer have mastered.

Thus, too, the equilibrium model of classical economics has inherent limitations for the understanding and mastery of the management and administration of natural resources. One of the few methodologically impressive attempts to apply symbolic logic to resource use was postulated mainly upon a relatively stabilized physical and cultural environment.[27] Although a type of stability or equilibrium may be part of the goal of a rationalized system of resource management, stability exists neither in the biotic nor the ecologic world, nor in the economics or politics of resources. It might be more fruitful to start with the presumption of an environment and an economy in flux, in a constant state of remaking and restabilization, if not in a continually unstable state. For this reason, too, the time schedule for achieving a readjusted and newly balanced resource policy can probably be more readily advanced by more eclectic techniques of analysis than sheer quantification.

On the political side, the overwhelming advantage of quantification in our era is that it lends confidence to judgments of quality about value-laden issues. On the administrative side, the foremost value of the computer, aside from routine operations, is to enable us to keep

---

[27] Walter I. Firey, *Man, Mind and Land: A Theory of Resource Use* (Glencoe, Illinois: Free Press, 1960).

policy abreast of the current situation through the use of constantly compiled data about the changing environment.

We face difficult philosophic issues in epistemologically relating quality and quantity, as well as ethical issues which go beyond semantics and statistics. But these issues have been faced by philosophers and scientists throughout history. Through further exercises in sheer precision we shall probably add less than we now expect to what will eventually prove fundamental to that discourse.

Assuming, however, that there will be substantial advances and advantages in precise mensuration and mathematical modeling, can it be anticipated that qualities will become increasingly measurable in quantitative terms and that, in effect, qualities will cease to be qualities and graduate—or descend—into the category of quantities? I think not. For, once we begin to measure qualities so faithfully, we merely increase our descriptive capacities and refine our manipulative skills, thereby accelerating the rate of change and extending the number of questions that may be raised about emerging phenomena. The range of value judgments which will then have to be made, and the burden of formulating policies in such an adumbrated world, will remain either undiminished or, more likely, will be immeasurably proliferated.

ON SOME MEANINGS OF "PLANNING"
William Petersen

More and more of the social processes of the twentieth-century world are "planned." Five-year plans, once restricted to the Soviet Union, have spread not only to other Communist states but to such diverse countries as India and Brazil. In the United States, which is featured in Communist propaganda as the last capitalist redoubt, the whole social-economic structure was altered by the government's response to the depression of the 1930's and to World War II. And city planners have extended their purview to regions and then, through local, national, or international agencies, to the whole physical-social environment of human life.

This seeming ubiquity of "planning," however, depends mainly on the application of a stylish word to modes of thinking and behavior that differ as radically now as they ever did.

> The word "planning" . . . has been known to cover the shooting of those who disagree, as under Stalin, or it may mean nothing more than consultation, making sure that all interested parties are properly informed, as in Ireland. It may involve the setting of targets with no compulsion to achieve them, but with certain inducements to try, as in France; or it may involve the setting up of targets with a large number of physical controls to back them up, as in India. And it can mean no more than the attempt to work out what consistency requires, to investigate what needs to be done if a given objective is to be achieved, as in England.[1]

Are there common elements that link these meanings and, if so, what are they?

---

[1] Maurice Zinkin, *Growth, Change and Planning: Economics of Developing Countries* (London: Asia Publishing House, 1965), p. 28. For a more complete and systematic comparison of planning in various countries, see Jan Tinbergen, *Central Planning* (New Haven: Yale University Press, 1964), pp. 104 ff.

When I delivered a preliminary draft of this essay to a Berkeley class in planning theory, their response was a curious combination of naïveté and cynicism. Of course, they told me, what they are doing is precisely the opposite of what planning was a generation ago and still is among certain middle-aged laggards. Everyone in the profession knows of this difference. But at the present time, while the new ideas are still evolving, they believe it would be unfortunate to distinguish the older and the newer schools too precisely, for the present lack of clarity affords practitioners a greater range of legitimate activity. So long as "planning" can mean almost anything, planners can both use the approbation the concept brings and avoid the limitations imposed by any single designation of function.

There is a parallel between the present ubiquity of "planning" and the conquest of eighteenth-century thought by "reason" or of nineteenth-century thought by "science." Darwin's *Origin of Species* excited the last major battle between the defenders and the opponents of scientific analysis. Intermediate disciplines underwent a painful and largely sterile self-analysis; half a century ago the *American Journal of Sociology,* for example, was obsessed with the question whether sociology is a science. Today, when we have the "science" of psychoanalysis, even the "science" of metaphysics, members of "the social sciences" seldom see any reason to question whether they are correctly subsumed under so broad a designation. The victory of "science" has been all but total—not over nonscience, but over the clear thinking that is based on meaningful distinctions, and thus over the efficient choice among policy alternatives.

DEDUCTIVE PLANNING

The average lay person is likely to understand "planning" in the deductive sense, following what is still the only dictionary definition. The planner draws up a blueprint on a flat surface or, in Latin, *planum;* and the design is completed before the first steps are taken toward its realization. "Planning results in blueprints for future development; it recommends courses of action for the achievement of desired goals." [2] This conception of planning was an extension to larger and more complex entities of the architect's design of a building or, less obviously, of the landscape architect's design of a park.[3] "The

---

[2] John C. Bollens and Henry J. Schmandt, *The Metropolis: Its People, Politics, and Economic Life* (New York: Harper, 1965), p. 278.

[3] Frederick J. Adams and Gerald Hodge, "City Planning Instruction in the United States: The Pioneering Days, 1900–1930," *American Institute of Planners Journal,* Vol. 31 (February, 1965), pp. 43–51.

Essence of City Planning is City Designing. . . . It is landscape designing in a larger phase." [4] The closest parallel to an architect's blueprint is the master plan; in the words of one practitioner, "the preparation and maintenance of the general plan is the primary, continuing responsibility of the city planning profession, . . . our most significant contribution to the art of local government." [5]

But the detailed design, a useful instrument to plan a building, is less appropriately applied to whole cities or national economies. In order to bring within its narrow compass the dynamic complexities of large units, the planner is forced to choose between a drastic restriction of his function and a narrowed perception of the matter he deals with. When he picks the second alternative, he usually evolves some monistic theory to reconcile his broad purpose with the limited realization possible to him. City planning, thus, is both the control of cities' growth by a "master" or "general" plan and also, by one common definition, the control merely of the city's physical elements—streets, parks, transportation, and the rest. The resolution of the contradiction is by what one analyst, after a study of more than a hundred master plans, termed "the physical bias"—the notion that the physical factors are the key determinants of all others. [6] In the words of Alfred Bettman, "The comprehensive master plan of city, town, region, state, by determining the appropriateness of place or location, and the program of urgency or priority, by determining the element of time, are instrumentalities for the creation of social values." [7]

A similar ecological monism plagues our theories about housing. A slum is ordinarily defined as a dense area of dilapidated structures and inadequate physical facilities, inhabited by those with economic, physical, mental, educational, or other disabilities. One way of perceiving the slum would be as a physical–social system whose parts interact: houses are dilapidated *because* of the residents' style of life, as well as vice versa. From such a perception, the remedial policies

---

[4] Crawford, in *City Planning* (1927); quoted in Charles M. Haar, ed., *Land-Use Planning: A Casebook on the Use, Misuse, and Re-use of Urban Land* (Boston: Little, Brown and Co., 1959), p. 45.

[5] T. J. Kent, Jr., *The Urban General Plan* (San Francisco: Chandler Publishing Co., 1964), p. 2.

[6] David Farbman, "A Description, Analysis, and Critique of the Master Plan," unpublished study cited in Paul Davidoff, "Advocacy and Pluralism in Planning," *American Institute of Planners Journal*, Vol. 31 (November, 1965), pp. 331–338.

[7] Quoted in Frederick J. Adams, "Recent Trends in Town Planning in U. S. A.," *Urban and Rural Planning Thought*, Vol. 1 (January, 1958), pp. 7–11.

that planners recommend would not be restricted to the physical plant. In fact, however, ameliorative actions to cope with some of the slum's social ills (more police to guard personal security, efforts to reinforce family stability) are considered reactionary, and since the New Deal American liberals have concentrated on improving the ecological setting. The myth of housing reform, as John Dean termed it some years ago, is the notion that if you remove slum housing you remove with it all the social pathologies of the slum. "It would be just as illogical to say that ills of slum areas are caused not by substandard housing conditions, but by the absence of telephone service, which also correlates with indexes of social disorders." [8] Perhaps it should be added, to avoid misunderstanding, that this is an argument not against public assistance for better housing, but against the monistic theories by which this policy is supported.

Planners of national economies, since they try to cope with a larger and more complicated system, are even more prone to squeeze their subject into manageable simplicity. America's program of aid to underdeveloped areas is based, at least in part, on a belief in crude economic determinism: ignoring the counter examples of Nazi Germany, the Soviet Union, and prewar Japan, our policy makers apparently believe that industrialization brings with it a full cornucopia of associated benefits, from greater democracy to mass culture. The notion that changes in the economic base determine the movement of the rest of the society, the noneconomic "superstructure," derives partly from Marx, but is not limited to Marxist analysts; all who use the term "economic development" to denote the precursor to general social advance follow this line of thinking.[9]

In an important sense, planning was foreign to Marx's thought. The doctrine that the inner workings of capitalism would lead to its inevitable collapse, and thus to the establishment of socialism, was what, to Marx, distinguished his "scientific" socialism from "utopian" varieties, which depended on man's rational will rather than

---

[8] John P. Dean, "The Myths of Housing Reform," *American Sociological Review*, Vol. 14 (April, 1949), pp. 281–288. According to the now famous —or, in the view of some critics, notorious—Baltimore study, lower-class Negroes who moved to better housing, compared with those who remained in the slum, showed few improvements in physical, mental, and social health. One need not accept every detail of this study's elaborate methodology to be struck by this conclusion. See Daniel M. Wilner *et al., The Housing Environment and Family Life: A Longitudinal Study of the Effects of Housing on Morbidity and Mental Health* (Baltimore: Johns Hopkins Press, 1962).

[9] See W. W. Rostow's "non-Communist manifesto," *The Stages of Economic Growth* (Cambridge: University Press, 1960).

economic laws as the main impetus to change.[10] Engels could not have been more contemptuous of the "crude theories" of his socialist predecessors:

> The solution of the social problems . . . the utopians attempted to evolve out of the human brain. Society presented nothing but wrongs; to remove these was the task of reason. It was necessary, then, to discover a new and more perfect system of social order and to impose this upon society from without . . . These new social systems were foredoomed as utopian; the more completely they were worked out in detail, the more they could not avoid drifting off into pure phantasies.[11]

In Marx's and Engels' own writings the nature of the socialist future is barely suggested, and this refusal to specify the society they advocated became a mark of orthodoxy among Marxists of all schools.

In the socialist parties of the West, the main, almost the sole, criterion of a socialist society used to be the public ownership of the means of production. In the 1951 Frankfurt Manifesto, which marked the postwar rebirth of the Socialist International, the emphasis shifted to "democratic planning" in a mixed economy as the basic condition of socialism. After more than a century of unremitting propaganda for nationalization, this was abandoned by the dominant factions in Germany's Social Democratic Party and Britain's Labour Party. As Crosland wrote in *Encounter,* paraphrasing the opening sentence of *The Communist Manifesto,* "A specter is haunting Europe—the specter of Revisionism." The "first and most obvious" reason for the new stance was that, when industries had been nationalized, "the reality proved rather different from the blueprints. Some of the anticipated advantages did not materialize, while certain unexpected disadvantages emerged." [12]

Although Lenin adhered to historical materialism in theory, in

---

[10] For an excellent discussion see Robert V. Daniels, "Fate and Will in the Marxian Philosophy of History," *Journal of the History of Ideas,* Vol. 21 (October–December, 1960), pp. 538–552.

[11] Frederick Engels, *Socialism, Utopian and Scientific* (New York: International Publishers, 1935), p. 36.

[12] C. A. R. Crosland, *The Future of Socialism* (London: Jonathan Cape, 1956), p. 316. A more telling break was made by Richard Crossman, a prominent Labour Party left ideologue and thus a frequent opponent of Crosland's. "What we describe as the Welfare State," Crossman wrote in 1956, "has been immensely successful and immensely popular, whereas nationalization has not changed the lives of the workers in the industries affected in the way they expected." Even more amazing was his conclusion: "Socialism cannot and should not be based on any particular economic theory." R. H. S. Crossman, *Planning for Freedom* (London: Hamish Hamilton, 1965), pp. 61–62.

practice he did not depend on the spontaneous reaction of the working class to "immiseration." The decisive element in taking power would be the Leninist party, a small, tightly knit organization of professional revolutionaries. And in 1917 the Bolsheviks took power not in order to create a specified kind of society but in order to have power. Apart from socialist slogans like "Production for Use," their program consisted in "Peace," "Land," "Bread," "Workers' Control" —rallying cries that reflected accurately enough what the Russian mass wanted, but in part contradicted Bolshevik principles and subsequent practice. When Lenin announced, the day after he established control, "We shall now proceed to construct the socialist order," [13] he could lead the government only to the destruction of the capitalist economy, a few catastrophic attempts at economic rationalization, and the eventual retreat to the mixed economy of the New Economic Policy. Planning in Russia did not really begin until 1928; its progenitor was not Lenin but Stalin.

The goals of the First Five-Year Plan, the methods by which they were to be achieved, and the complex allocations of human and material resources were spelled out in three large volumes, which proved to be a grossly inadequate tool for guiding the whole economy and society through a period of forced rapid change. The details included in subsequent plans were so reduced that by the time of the Fourth Five-Year Plan, the first after the defeat of Nazi Germany, the general directive consisted of a single pamphlet. Its function was to incite enthusiasm for the broad goals of restoring prewar levels of production, overtaking and surpassing the capitalist nations, and moving from socialism to communism. The actual guide to the development of the economy was given in the quarterly plans, which were loosely drawn and adjustable to unforeseen circumstances. More recently, a number of economists in the Soviet Union and other Communist countries, trying to overcome their ideological encumbrances, have advocated that the half-surreptitious use of the market be recognized and extended.[14] For if the market is admitted as a legitimate tool of a planned economy, then Western economic theory, that powerful analytical instrument, can be used to improve the quality of the

---

[13] Merle Fainsod, *How Russia Is Ruled* (Cambridge: Harvard University Press, 1957), p. 84.

[14] See several of the essays in Gregory Grossman, ed., *Value and Plan: Economic Calculation and Organization in Eastern Europe* (Berkeley and Los Angeles: University of California Press, 1960). A good article on the trend in the Soviet Union is Leon Smolinski and Peter Wiles, "The Soviet Planning Pendulum," *Problems of Communism*, Vol. 12 (November–December, 1963), pp. 21–34.

planning. Even in the original home of planned economy, "planning" denotes mainly a propaganda façade, together with a rationalized but relatively loose pattern of economic development.

UTOPIAN PLANNING

A deductive plan, when applied to a large, complex, dynamic system, tends thus to break down into more manageable parts, or to be broken down by a systematic distortion of reality. Alternatively, a planner with ideas too grandiose for his practical world may take off into fantasy. Though planning ordinarily connotes the introduction of greater rationality into social policy, it can also mean the abandonment of reason. My impression is that utopian thought has undergone a revival during the past decade. Not only have such works as the Goodmans' *Communitas* and Howard's *Tomorrow* come out in new editions, but one of the better recent texts on urban sociology includes a sympathetic chapter on "The Visionary"; [15] and the dean of an important school of city planning has suggested that "planners ought to recognize the value of utopian formulations in the depicting of the community as it might be seen through alternative normative lenses." [16] To analyze utopianism is not, as it might have seemed some years back, beating a dead jackass.

1. A utopian formulation is the statement of a purpose so broad, so lofty, that if it means anything at all it denotes an unattainable goal. "The nurture of life is the main aim of collective endeavor"; [17] the goal of planning is "substantial gains in human happiness." [18] Such consciously unrealistic aims are supposed to be useful as a goad; in Mumford's words, while adaptations will be necessary as the plan "encounters the traditions, the conventions, the resistances, and sometimes the unexpected opportunities of actual life," it would be a mistake to anticipate these necessary changes. Only "overbold" plans will "awaken the popular imagination: such success as totalitarian

---

[15] Leonard Reissman, *The Urban Process: Cities in Industrial Societies* (New York: Free Press of Glencoe, 1964), chap. 3.

[16] Martin Meyerson, "Utopian Traditions and the Planning of Cities," *Daedalus*, Winter, 1960, pp. 180–193. Cf. Dahl and Lindblom: "As models, utopias . . . indicate directions in which alternatives to existing reality might be looked for, . . . help one to focus on long-run goals, . . . function as aids to motivation." Robert A. Dahl and Charles E. Lindblom, *Politics, Economics, and Welfare* (New York: Harper, 1953), p. 73.

[17] Lewis Mumford, *The Culture of Cities* (New York: Harcourt, Brace and Co., 1938), p. 377.

[18] David Riesman, *Individualism Reconsidered and Other Essays* (Glencoe, Illinois: Free Press, 1954), p. 73.

states have shown in their collective planning has perhaps been due to their willingness to cleave at a blow the Gordian knot of historic resistances." [19]

Comment: Whether holistic utopian schemes act as a leaven to workaday dough is an empirical question, which to my knowledge has never been seriously examined. Some important points suggest that the thesis is invalid. Frequently, perhaps typically, the utopia is seen not as an incentive but as an alternative to social reform. Such aims as "a chicken in every pot," or "a car in every garage," which *can* be achieved, are, in Riesman's perspective, no more than ideological pressures of business enterprise.[20] In the eyes of the true utopian, anything that is feasible is by that fact not worth attempting; every revolutionary party has devoted its major effort to fighting reformists. Both extremes of the political spectrum attack the center, its relatively modest aims and moderate means of achieving them; the wild charges that the radical right and the extremist left make against each other are mutually supportive.[21] Among both city planners and Soviet administrators, the contrast between The Plan and actual planning induces chronic frustration and a consequent search for a scapegoat—persons or institutions that have not fully cooperated in furthering The Plan. One need hold no brief for either real estate boards or Western governments to find this devil view of social processes naïve and, in a totalitarian context, the prelude to terror. Very often the details that are omitted from the utopian's broad vista relate to the attitudes and desires of the people who will be affected. Those who oppose The Plan, then, are depersonalized into Mumford's "historic resistances," while those who impose it acquire special license from the splendor of the world they are creating.

2. An even more characteristic stance of the utopian than his demand for the vague and unattainable Good is his denunciation of the equally vague Evil—"the system," "the establishment," "the power elite," and so on. Mumford provides a convenient example: "the debris of these dying systems" includes, by his broad compass, "Orthodox Christianity, Protestantism, individualistic humanism, capitalism, humanitarianism, and libertarianism." [22] The utopian is

---

[19] Mumford, *op. cit.*, p. 380.

[20] Riesman, *op. cit.*, p. 73.

[21] As a migrant from the more consistently democratic East Coast, I have been struck by the way this principle operates in California politics. The major parties in this state are seriously infected with utopian factions. Each faction, in its own manner, would like to go beyond the orderly processes of legal change—and is reinforced in the conviction that this is necessary by those seeking quasilegal methods to move in the contrary direction.

[22] Mumford, *op. cit.*, p. 378.

"for" planning *tout court;* "the irrational and planless character of society must be replaced by a planned economy" in order to achieve "self-realization for the masses of the people." [23] Apart from planning there is, in this perspective, no order in society. Capitalist production is, as Marx pointed out, "anarchic." Any unplanned city is "this chaos of congestion, this anarchy of scatteration." [24] Or, alternatively, whatever the consistent patterns in the unplanned social world may be, their effect is totally deleterious. *Any* plan is better than no plan at all.

Comment: It is a remarkable procedure to begin a more rational control of society by blithely dismissing the whole content of the social disciplines. If outside the planned sectors there is nothing but anarchy and chaos, then economists, sociologists, and political scientists have certainly been adept at inventing the recurring patterns they study. For Marx, the "anarchy" of capitalist production was no more than a journalistic gloss on a life's work in trying to establish the "laws" of capitalist development. It is true that the laws of social behavior are not as absolute as he depicted them, but, inadequate as our knowledge of society may be, it is a better basis for control than a self-imposed ignorance.

The market, to return to that recurrent example, is an efficient mechanism for distributing commodities in order to achieve the maximum over-all utility. One may object that this is not the only social goal to be sought, that labor and land are not commodities and should not be subjected to market conditions, that many so-called markets are half-disguised monopolies, and so on; but the point remains that the market is a good tool for certain important purposes. Even in Communist countries it has been deemed necessary to introduce the market principle, to the degree that the utopian heritage of Marxian thought permits.

The tendency of planners to depreciate all social regularities is important enough to make another example appropriate. In *American Skyline,* Tunnard and Reed demonstrate with a wealth of fascinating detail that "definite forces" (with city planning among the least of them) have molded the physical patterns of urban America; but they do not spell out the implications of this fact for present-day efforts to control these forces. As in the work of many other planners (the senior author is Professor of City Planning at Yale), the main burden

---

[23] Erich Fromm, *Escape from Freedom* (New York: Rinehart, 1941), p. 272.
[24] Victor Gruen, "How to Handle This Chaos of Congestion, This Anarchy of Scatteration," *Architectural Forum*, Vol. 105 (September, 1956), pp. 130–135.

of their excellent study is to contrast the "city efficient" with its un-planned and therefore inefficient counterpart.[25]

3. An engineer operates on the assumption that if he lowers cost he *generally* sacrifices quality; if he has to cut the weight of his material, he *usually* loses some of its strength; and so on. The expectation of the typical social engineer is the contrary. Family sociologists, almost to a man, are deeply concerned about family instability, yet few seem to recognize that this is in part a consequence of the social trends they support. In a society where women have minimal legal rights and no possibility of earning their own living, wives are less likely to consider divorce no matter what the provocation; and in a society where young men routinely follow their fathers' occupation, with no opportunity to rise above it, the link between generations is apt to be strong. Women's rights, social mobility, and family stability are three elements of an interrelated system. To maximize any one of these goals is a feasible project, but an attempt to realize all three goals completely is a form of utopianism to which all too many of us succumb in our enthusiasm for particular projects.

A recurrent dilemma for democratic city planners is that their professional efforts to structure space into functional units may result in the hierarchical ordering of the persons occupying the space. Perry's neighborhood principle, designed to create a "community in which the fundamental needs of family life will be met more completely," was based on the doctrine that the elementary school should be the neighborhood's focus.[26] For a period his thesis was generally accepted—until "de facto segregation" became an issue; and then some began to assert that "the 'neighborhood unit' is an instrument for segregation." [27] It is not necessary here to argue the comparative worth of reinforcing social units intermediate between the family and the city, and of facilitating the movement of ethnic minorities into the general society. The point is, by our values both goals are desirable, but they are partly incompatible. A planner who propounds one or the other course should at least be aware that its cost includes a relative sacrifice of the other. In particular, he should try to avoid advocating both at the same time. In a recent article, Frieden (in my

---

[25] Christopher Tunnard and Henry Hope Reed, *American Skyline: The Growth and Form of Our Cities and Towns* (New York: Mentor, 1956).

[26] Clarence Arthur Perry, *Housing for the Machine Age* (New York: Russell Sage Foundation, 1939).

[27] Title of an article by Reginald R. Isaacs in *Journal of Housing*, Vol. 5 (August, 1948), pp. 215–219. See also Nathan Glazer, "The School as an Instrument in Planning," with a "Comment" by John W. Dyckman, *American Institute of Planners Journal*, Vol. 25 (November, 1959), pp. 191–199.

opinion, one of the more perceptive writers on city planning) at least implicitly accepts the commonplace that a prime fault of metropolitan structure is that the central city, which carries the main burden of providing services for the whole region, has lost much of its tax base. However, a policy of rebuilding central cities by attracting some of the more prosperous back to them, "far from assisting low-income families, . . . is a threat to their welfare." [28] As before, the intent here is not to argue for one course or the other but only to point out that they are incompatible, and thus to raise the question whether the welfare of the poor is enhanced if they are encouraged to remain segregated in central cities, even if in public housing.

Comment: While often only one of several related goals can be fully realized, sometimes the interaction is mutually supportive. Perhaps the best-known case is what Myrdal termed "the principle of cumulation." "White prejudice and discrimination keep the Negro low in standards of living, health, education, manners and morals. This, in its turn, gives support to white prejudice. White prejudice and Negro standards thus mutually 'cause' each other. . . . If either of the factors changes, this will cause a change in the other factor, too, and . . . the whole system will be moving in the direction of the primary change, but much further." [29] It would be useful to distinguish, other than *ad hoc,* between the two broad types of equilibrium. Although many have used Myrdal's example as a general model, it is in several respects a special case. The inferiority of Negroes is not innate, and can thus respond fully to an ameliorative environment. Consider, in contrast, the "social problem" of the aged; all too much written about, it is based on a refusal to accept the fact that here the physiological distinction *is* crucial. If the Negro's situation improves, the persistence of irrational prejudice must combat a value system so strong that Myrdal termed it "the" American creed—that is, the belief in equality of opportunity irrespective of race. In contrast, if all the real problems of urban life were to be solved—traffic congestion, smog, slums, excessive crime, and so on—the city planners who now plan against the city could feed their prejudices from the strong irrational stream, dating from Washington and Jefferson, that colors most of our urban policy. [30]

---

[28] Bernard J. Frieden, "Toward Equality of Urban Opportunity," *American Institute of Planners Journal,* Vol. 31 (November, 1965), pp. 320–330.
[29] Gunnar Myrdal, *An American Dilemma: The Negro Problem and Modern Democracy* (New York: Harper, 1944), pp. 75–76.
[30] See Morton and Lucia White, *The Intellectual versus the City: From Thomas Jefferson to Frank Lloyd Wright* (Cambridge: Harvard University

INDUCTIVE PLANNING

Thus far, in the effort to bring order into the variety of meaning given to the word "planning," we have considered two main types, deductive and utopian. The third type, inductive planning, attempts to coördinate public policies in several overlapping economic and social areas. In the words of Myrdal, "Coordination leads to planning or, rather, it *is* planning as this term has come to be understood in the Western world. . . . The need for this coördination arose because the individual acts of [government] intervention, the total volume of which was growing, had not been considered in this way when they were initiated originally." [31] Planning in this sense is "pragmatic and piecemeal and never comprehensive and complete"; plans usually constitute "compromise solutions of pressing practical issues." [32]

It is no criticism of this kind of policy making per se to ask the question: If it is coördination, if it is administration, why then is it useful to term it "planning"? [33] We are like Molière's *bourgeois gentilhomme* when he discovered he was speaking prose: for more than forty centuries we have been planning without knowing it.[34] The answer to the question, presumably, is that "planning" is a special kind of prose, different indeed from the classic iambs of blueprints and the free verse of utopians, but different also from the prose of earlier bureaucratic operations. Those who have been working for the past dozen years in reaction against the two earlier traditions have produced enough to start us thinking how these several strands can be brought together. In such a synthesis, we would begin with the most apparent elements of the developing new consensus:

1. Planning concerns process and not state; it pertains not to some idealized future but to the mode of moving from the present.

Press and MIT Press, 1962); Henry Nash Smith, *Virgin Land: The American West as Symbol and Myth* (New York: Vintage Books, 1957).

[31] Gunnar Myrdal, *Beyond the Welfare State: Economic Planning and Its International Implications* (New Haven: Yale University Press, 1960), p. 63.

[32] *Ibid.*, p. 23.

[33] In a rather good book on economic development, written by a socialist administrator, the item under "Planning" in the index reads as follows: "*See* Government (administrative functions); Investment criteria; Price mechanism." W. Arthur Lewis, *The Theory of Economic Growth* (Homewood, Illinois: Richard D. Irwin, 1955), p. 448.

[34] "In this sense there has always been economic planning . . . in the days of the Pharaohs of old, in ancient Greece, in the Roman Empire. . . . There has always been town planning, planning in relation to great public works." Ferdynand Zweig, *The Planning of Free Societies* (London: Secker and Warburg, 1942), p. 11.

2. The traditional concentration on the physical plant (city planners) or on the economy narrowly defined (national planners) is a false emphasis. "A plan for the physical city has utility only as a step in a means-end continuum that causally relates the artifactual city to the social-economic-political city." [35]

3. As a corollary, the relevant skills are not—or not exclusively —architecture, landscape architecture, sanitation engineering, and the like (city planners), or economics, economic geography, formal demography, and the like (national planners), but rather the full range of social sciences, especially some of the newer emphases or subdisciplines: the study of political determinants of social stability and of decision making; systems analysis, with the computer as the fundamental tool and model building as a routine step; sociology so far as this departs from the Chicago ecological tradition and concentrates on the social setting of ethnic relations, poverty, education, labor-force recruitment, and similar policy fields.

4. An attempt to realize a broad, over-all plan (or, in the new language, to move forward on a broad front) is likely to be successful only if a "middle-range bridge" is built of specific functions assigned to particular agencies.[36] In a history of planning from this point of view, five-year plans would get as much attention as phlogiston in a history of science; and the analysis of city maps by a Reps would be amply supplemented with a McKelvey's account of the rise of municipal bureaus.[37]

5. The choice among competing goals, a key problem that is beyond ultimate solution, can sometimes be reduced to a manageable scale by extending to new uses the principles of the market and of cost accounting. For example, the so-called Victoria Line of the London underground, though manifestly needed, was ruled out because it could not "pay its way." But if one brought into the calculation the secondary benefits of relieving traffic congestion and saving the time of commuters, the investment could be shown to have a return of

---

[35] Melvin M. Webber, "The Prospects for Policies Planning," in Leonard J. Duhl, ed., *The Urban Condition: People and Policy in the Metropolis* (New York: Basic Books, 1963), pp. 319–330.

[36] Martin Meyerson, "Building the Middle-Range Bridge for Comprehensive Planning," *American Institute of Planners Journal*, Vol. 23 (Spring, 1956), pp. 58–64.

[37] John W. Reps, *The Making of Urban America: A History of City Planning in the United States* (Princeton: Princeton University Press, 1965); Blake McKelvey, *The Urbanization of America, 1860–1915* (New Brunswick: Rutgers University Press, 1963). McKelvey's work is the only one I know of that discusses the muckrakers (the utopians of that era) against a background of rising professionalism in city government and related social agencies.

more than 10 per cent; [38] and on this basis the project was started. This type of analysis is most useful when there is a consensus on the social benefits to be attained, and when they can be quantified without too great a strain on credibility.

There are several other fundamental areas that need further clarification, but I shall restrict myself to two: facts and evaluation.

Facts: Wesley Mitchell, who, like all the institutional economists, was a forebear of the American social planners of today, believed that tested factual knowledge is a prerequisite to adequate policy making. On the face of it, this may seem to be a reasonable point of view, but it was widely attacked as visionary and reactionary.[39] Indeed, our ignorance in many areas is so great that if we demanded an adequate empirical base for planning, in many cases we could hardly begin to plan. One list of "research needs" for city planning runs to sixty-three items, and includes everything from "why people move to the suburbs" to "how to devise a system of metropolitan government which is expandable with its area growth." [40] Moreover, the information that does exist is seldom available in a coördinated, usable up-to-date, form to the agencies that need it. One important "middle-range bridge" would be intelligence centers to process facts and provide inventories and forecasts.[41] If, in the meantime, the authorities intervene in citizens' lives with "a healthy pragmatism," "an appropriately experimental approach in the face of many unknowns," is this altogether "on the positive side"? [42] The utopian has an auto-

---

[38] C. D. Foster and M. E. Beesley, "Estimating the Social Benefit of Constructing an Underground Railway in London," *Journal of the Royal Statistical Society,* Series A, Vol. 126, Part 1 (1963), pp. 46–78.

[39] Forest G. Hill, "Wesley Mitchell's Theory of Planning," *Political Science Quarterly,* Vol. 72 (March, 1957), pp. 100–118, at n. 24. Lynd asserted that the research under Mitchell's direction at the National Bureau of Economic Research, since its purpose was to establish empirical relations in the world as it exists, had "a general bias in favor of the going system." Robert S. Lynd, *Knowledge for What? The Place of Social Science in American Culture* (New York: Grove Press, 1964), p. 121.

[40] Emil J. Sady, "Notes on Research Needs, Compiled from the Conference Records," National Conference on Metropolitan Problems Held at Kellogg Center, Michigan State University, East Lansing, Michigan, April 29–May 2, 1956, *Proceedings* (New York: Government Affairs Foundation, 1957), pp. 87–89. For an intelligent variation on that common theme, "more research is needed," see *A Report on Planning, Policy-Making, and Research Activities* (Washington, D. C., Resources for the Future, 1961).

[41] Melvin M. Webber, "The Role of Intelligence Systems in Urban-Systems Planning," *American Institute of Planners Journal,* Vol. 31 (November, 1965), pp. 289–296.

[42] Harvey S. Perloff, "Social Planning in the Metropolis," in Duhl, ed., *op. cit.* in note 35 above, pp. 331–347, at p. 331.

matic response to this moral dilemma, but what is it for the democratic empiricist?

If planning is seen as a process, the empirical base required is not merely isolated facts, but facts in correlated moving systems, or projections. If a model is constructed, for example, showing how land use, transportation, and human behavioral patterns interact, it is possible to build an efficient highway relative to the other two factors and, if the system is projected into the future, to provide a continuing basis for public decisions on these matters. Or so we are told in an important article.[43] I gladly accept what I take to be an advance in an area about which I know nothing, but I am less certain that the method can be extended to matters about which I am more knowledgeable.

Population forecasts are a good test of the general utility of projections: the data are excellent, and information on the trend in the number of people affected is basic in virtually any kind of planning. But the record of success has not been one to inspire confidence. The United States Census Bureau abandoned the use of the word "forecast" for "projection" and then began to use what are termed "illustrative projections."[44] For local areas, where migration is often an important determinant of population growth, many projections are hardly even illustrative—except of our ignorance. One important reason why population projections often fail is that they are automatically tested: the hypothesis never is that this transportation system is the most efficient, or the cheapest, of all the systems that might have been built. Another reason is that what we know of human behavior pertains mostly to that portion or it defined by Economic Man; the motives that lead parents to decide, for instance, whether to have a third child are still largely a matter of speculation.

We have spoken of "facts" as though this were a nonproblematic term. "Factual statements and their analysis invariably reflect the values of their makers, if only in the importance attached to them or the sequence in which they are studied."[45] It is a measure of the harm that simplistic positivists have done to the social disciplines that

[43] Britton Harris, "Plan or Projection: An Examination of the Use of Models in Planning." *American Institute of Planners Journal,* Vol. 26 (November, 1960), pp. 265–272.

[44] A good critique, even though its examples are out of date, is Harold F. Dorn, "Pitfalls in Population Forecasts and Projections," *Journal of the American Statistical Association,* Vol. 45 (September, 1950), pp. 311–334. A more general discussion is given in William Petersen, *Population* (New York: The Macmillan Co., 1961), chap. 11.

[45] Paul Davidoff and Thomas A. Reiner, "A Choice Theory of Planning," *American Institute of Planners Journal,* Vol. 28 (May, 1962), pp. 103–115.

it is still necessary to make this kind of elementary rejoinder. Is the word "invariably" appropriate? Does the statement that the population of the United States in 1960 was almost 180,000,000 reflect in any sense, to any degree, the values of the Director of the Census? People have many values, sometimes in conflict; which ones are reflected in factual statements? One important value system is to strive to tell the truth; when a scientist thoroughly imbued with the ethics of his calling reflects this value, does he give "true" facts? In short, those who wander into the social disciplines to get some guidance on how to plan should acquire an epistemological sophistication beyond the level of simple antipositivism.

Evaluation: When planning introduces greater rationality into an area of life, planning is in this respect akin to science. A scientific proposition is one that can be proved fallacious by comparing it with the empirical world; a scientist constantly shuttles between his hypothesis and his data, between the model he constructs and the pattern he observes. The analogous process is the evaluation of plans, which would be equally routine if planning were scientific in the same sense. On the contrary, evaluation is atypical and, if made, often inadequate.

With respect to utopian plans, what can evaluation signify? If the purpose of the plan is a paraphrase of Riesman's "substantial gains in human happiness," it will never be possible to determine in any precise sense whether this has been achieved. Since a utopian goal keeps receding as we approach it, the goal cannot be used to measure progress along the road toward it. Nor has the supposed proximate value of a utopia, to supply the incentive for achieving realizable goals, ever been more than supposed. The evaluation of utopian plans, then, is ruled out as impossible or meaningless, and we must remember that many nonideological plans contain a utopian element.

Evaluation of inductive plans is obviously difficult, since the measuring stick and the thing to be measured constantly interact. The most efficient type of operation when it is feasible, a feedback system, means that evaluation in one sense takes place constantly,[46] but not the determination of whether any operation is successful in realizing a given fixed goal.

Let us restrict ourselves here to the relatively simple case of deductive planning, since even that involves problems of what is meant by "evaluation." A schematic typology may help in differentiating the situations to be discussed:

---

[46] Cf. Britton Harris, "New Tools for Planning," *American Institute of Planners Journal*, Vol. 31 (May, 1965), pp. 90–95.

Actual change

		no	yes
Planned change	no	1. "Natural" stability	2. "Natural" change
	yes	3. "Unsuccessful" planning	4. "Successful" planning

1. "Natural" stability, one might think, is irrelevant to a discussion of planning. But if no policy has been initiated only because the process is seen as "natural" (that is, beyond the realm of policy), that definition of it may change. To the degree that planning connotes an attempt to establish rational control over social processes, it results in a contraction of "nature." For instance, when children were welcomed in Western countries as "gifts of God," fertility was regulated by institutional patterns (such as the postponement of marriage from puberty to some "appropriate" age), but no type of birth control was considered legitimate. The "modernization" or "westernization" of the world usually means the extension of rational control; hence no present designation of "nature" can give a clear indication of what it will be in the future.

The "lack" of policy may sometimes be a conscious, deliberate decision not to act. The existence of the Tennessee Valley Authority and the nonexistence of a Missouri Valley Authority are both consequences of policy decisions; if the first represents an important influence of the federal government on one region, the second does also. Or, as another example, perhaps the most effective way to inhibit the overrapid growth of cities in underdeveloped areas is to cut down the overrapid growth of the general population; governments that attempt to control the first but do nothing about the second are in the most meaningful sense acting in both fields. Logically, then, we ought to include some types of nonplanning in a survey of planning, but with such a definition the subject would become almost coextensive with the social world. In practical terms, decisions not to act must be omitted from a useful definition of planning.

2. "Natural" change, for the true utopian, should be a null category. Is there any segment of society that could be left unplanned in order to attain "self-realization for the masses"? A meaningful goal, however, ordinarily pertains to only a small segment of social life, and its success depends mainly on how it fits in with the larger, more

significant, unplanned sector. Success in planning depends on our knowledge of the whole society, which is hardly adequate.

3. "Unsuccessful" planning, in theory, is of fundamental importance. The true test of a plan is its failure, which is parallel to the negation of a scientific theory. For it is principally by proved deficiencies in specific plans (or theories) that planning (or knowledge) can advance. To carry through the parallel, the test of success ought to be made independently, not by the man with a personal stake in demonstrating success.

4. "Successful" planning means, at a simplistic level, planned change that is congruent with actual change; the question is when may we reasonably substitute "the determinant of" for "congruent with." In any but the least complex, most concrete social policies, establishing a causal link between purposive acts and their possible consequences requires a full-scale analysis. If a policy has multiple purposes, some may be realized and others not. Is the policy then successful?

Such points can be made better in terms of examples. The Venezuelan government's housing agency supervised the construction of apartments in Caracas to accommodate an estimated 180,000 persons. In mechanical terms the project was completed: 97 fifteen-story buildings were erected at a total cost of some $200,000,000. But 4,000 families invaded the apartments and lived there illegally, while others squatted in the community facilities or in shacks built on the project site. Unpaid rents totaled $5,000,000; losses in property damage amounted to $500,000 a month; delinquency and crime rose appreciably; tenant associations, controlled by agitators, impeded control measures and built up a potentially explosive atmosphere. An international interprofessional team, called in to study the situation, recommended that "the government should suspend all construction of superblocks until there exists a defined housing policy related to the economic and social development of the country and within a process of national planning and coördination. It was found that the massive construction programs in Caracas had served to attract heavy migration to the city from rural areas and, therefore, severely intensified the housing problem in the capital." [47]

Sometimes the effect of planning is not less but greater than was envisaged. The most important effects of city planning in the United States have been on construction not included in specific plans, much of

---

[47] Eric Carlson, "High-Rise Management Design Problems as Found in Caracas Studied by International Team," *Journal of Housing*, Vol. 16 (October, 1959), pp. 311–314.

which has reflected, sometimes as a kind of caricature, the planners' dicta of earlier years: the superlock, the abolition of the gridiron pattern, the stress on low density, or what is now called "scatteration." "A generation ago most experts saw great crowded cities as a destructive anachronism, to be drastically altered by some form of decentralization. But suburban sprawl is clearly no solution, and the big centers have survived despite decay and mounting congestion. Now the new generation of experts and critics tends to glorify the economic and cultural virtues of the Great City, scorning decentralization in any form." [48] It would be an exaggeration, but one with more than a germ of truth, to say that the plan of one generation becomes the social problem of the next. The "success" of planning often becomes an embarrassment.

As a final example, consider that thoroughly worked theme, Britain's new towns. The project, a discerning British town planner tells us, was "mainly successful," and in this opinion he joins in "the almost universal support" now given the project in Britain. The towns exist and are the homes of thousands. The Conservative Political Centre adjudged the new towns "a financial success"; and *Socialist Commentary*, published by a group well to the left of the Labour Party, called them "a striking financial success." [49] But surely the program was not instituted to make money; it was started, on the contrary, because the operation of the market system led to a number of social ills that the project was intended to ameliorate. To judge whether it was successful in this sense, we must look first at the goals with respect to the large cities from which the people were transferred, and then at the new towns themselves.

Perhaps the most authoritative and succinct expression of the first goal is the statement of purposes of the Barlow Commission, out of whose report the new-towns program developed. The Commission's assignment was:

to inquire into the causes which have influenced the present geographical distribution of the industrial population of Great Britain and the probable direction of any change in that distribution in the future; to consider what social, economic, or strategical disadvantages arise from the concentration of industries or the industrial population in large towns or in particular areas of the country; and to report what remedial measures, if any, should be taken in the national interest.[50]

---

[48] Catherine Bauer Wurster, "Framework for an Urban Society," in *Goals for Americans* (New York: Prentice-Hall, 1960).

[49] John Madge, "The New Towns Program in Britain," *American Institute of Planners Journal*, Vol. 28 (November, 1962), pp. 208–219.

[50] Royal Commission on the Distribution of the Industrial Population, *Report* (London: H. M. Stationery Office, 1940; Cmd. 6153).

The Commission inquired, considered, and recommended; and their policy *was* a success in the minimal sense that the new towns exist. But have "the social, economic, or strategical disadvantages" of industrial concentration been mitigated? In particular, has London stopped growing, or has it even grown less than it would have without the plan? In the opinion of Peter Self, "all available evidence suggests that the drawbacks to the indefinite growth of the conurbations, and particularly of London, are in fact much greater" than before World War II.[51] I am in process of trying to test his conclusion with the most recent available data, which seem to indicate that it still holds, particularly if one takes into account what might have been done in the metropolitan areas with the vast amounts of energy, money, and skill that were used to move factories and people out of them.

The intention with respect to the new towns can be stated best, perhaps, by citing the final report of the Reith Committee.[52] Beyond the facts that the towns exist and that they are paying their way, what is there that could be denoted a success? "There should be a full latitude for variety and experiment," the Reith Committee declared, with no "standardized pattern of physical or social structure." Most commentators, including those who, like Madge, are inclined to take a friendly view of the program, are disappointed that it has done so little to disturb the physical drabness of Britain's old towns and suburbs. The "optimum normal range" of population was given as "30,000 to 50,000 in the built-up area," with a total of 60,000 to 80,000 in the town plus its immediate environs. The question is not whether these limits will be breached, but when.

The Reith Committee was willing to congratulate itself on its realism in seeing that social classes still exist in Britain ("some would have us ignore their existence"); but the notion that the provinciality of the provinces can be legislated out of existence is no less strange. "So long as social classes exist, all must be represented in [the new town]. A contribution is needed from every type and class of person; the community will be the poorer if all are not there." The factory workers are accommodated with their factories; but if their sons, taking advantage of Britain's new emphasis on higher education for a larger proportion of the population, seek to rise to different skills, will not many of them be induced to move to the centers of British civilization?

The new towns represent the most massive and important effort

---

[51] Peter Self, *Cities in Flood: The Problems of Urban Growth* (London: Faber and Faber, [1957]), p. 27.

[52] New Towns Committee, *Final Report* (London: H. M. Stationery Office, 1946; Cmd. 6876).

within the framework of a dynamic urban-industrial system to reverse the trend toward greater urbanization and emphasize the "community" and "harmony" of pre-industrial society. Sometimes these are offered not merely as superior goals but as the aims of planning altogether: planning is *defined,* thus, as a means of directing change "toward the ultimate objective of orderly and harmonious community processes," [53] or as "methods and techniques to coördinate and bring into harmony" the uses made of land and the structures on it.[54]

In this view, harmony equals stagnation, lack of progress. Change takes place because a disjunction arises among the parts of an interrelated system; and a planner who wants to bring about more or different changes cannot dispense with this motive power. Only in the final stage of development, the utopia in which no ills remain, is harmony an entirely useful characteristic. But in utopia the planners will be unemployed.

---

[53] Noel Gist and L. A. Halbert, *Urban Society* (New York: Thomas Crowell, 1956), p. 480.
[54] Mary McLean, ed., *Local Planning Administration* (3d ed.; Chicago: International City Manager's Association, 1959), p. 10.

RESOURCE QUALITY: NEW DIMENSIONS
AND PROBLEMS FOR PUBLIC POLICY
Michael F. Brewer

The contemporary emphasis on highway aesthetics, the growing awareness of the problems of water pollution, our increased apprehension of the consequences to human health of smog, all have given rise to a broad public concern with the quality of our environment and its component resources. This concern has stimulated efforts to better understand the problems faced by users of particular resources, as well as by the general public, and to undertake corrective measures. This new sensitivity to quality poses new problems for public resources agencies and requires new dimensions of public policy.

Public policy in this area to date has not included prototype programs or legislation. The record of limited achievement has occasioned charges of major shortcomings in the programs of the federal agencies concerned. I suggest that these accusations are based on misconceptions. Federal policies and programs for natural resources operate through a complex network of organizations. The attempt to base public programs for resource management on our understanding of an ill-defined goal of quality imposes a heavy burden on the federal agencies, which are slow to respond because of their very complexity.

Thus far, the central concept of resource quality has eluded serious attempts at definition in the increasingly voluminous literature on the subject. For our present purposes, resource quality refers simply to attributes of a natural resource that are not identified by a generally accepted label and subject to a generally accepted system of measurement. It is the residual of a resource's properties. Public concern with natural resources usually varies directly with the extent to which technological insight suggests that the resource has potentialities for multiple uses.[1] If its productivity in those uses remains undiminished

---

[1] A frequent manifestation of concern with resource quality is a proposed method of classification that distinguishes between qualities. Soil classification

over a given time interval, and allowing for any change in general technology, it frequently is concluded that there has been no loss in quality. A loss in productivity of the resource for one or more of its uses amounts to a qualitative degradation.

A resource may have a single use, several individually specified uses, or an open-ended class of uses that are not specifically designated but simply referred to as future uses. The term "quality" may be applied also to an individual resource or to a cluster of several resources, as is done for the quality of environmental resources.

In what way does resource quality enter the domain of public policy, and what are its implications for the programs of resource agencies? I propose to identify the requirements imposed on the federal resource agencies by considerations of quality and to suggest ways in which these agencies might accommodate such considerations more effectively in their programs and policies.

The structure and authority of federal natural resource management practices have been dictated by the problems with which the agencies have had to cope rather than by any general theory of public administration. Their present policy has evolved from a long and diverse administrative heritage. Organizations, like individuals, become sensitized by past experience, and it is useful to know the pattern of past conditioning in order to identify the bottlenecks that may be encountered as they respond to new problems. For our purposes the chronology of the federal agencies can be traced by identifying three periods, or eras, and then by examining the most recent transition in policy.

THE TRANSITION OF OBJECTIVES

The three eras are abbreviated versions of those identified by Clawson and Held as "acquisition," "disposal," and "management." [2] The acquisition era commenced with the territorial arrangements made with the original thirteen states in 1779. Its most dramatic episodes were the Louisiana Purchase and the purchase of what is now southern Arizona from the Mexican government, in the first half of the nineteenth century. The resources were lands with accompanying minerals and timber, and the public agency executing the acquisition on behalf of the United States was its Land Office. After 1850, terri-

schemes exemplify this. Different classification systems have accompanied increases in technical insight into soil-plant relationships and cultivation.

[2] Marion Clawson and Burnell Held, *The Federal Lands: Their Use and Management* (Baltimore: The Johns Hopkins Press, 1957), pp. 15–44.

torial addition was negligible except for the purchase of Alaska from Russia in 1867.

The disposal era represented an effort by the federal government to develop agencies, legislation, and policies for a very different objective. Even while the United States was acquiring large blocks of western land, it was disposing of public land. The granting of private title to the public domain was in response to political pressures, economic demands, and prevailing social philosophies: the Jeffersonian hypotheses regarding the superior social responsibility and political stability of the freeholder; Hamilton's concept of the public domain as the national specie; and the mounting pressure for preemption by western settlers.

The Homestead Act of 1862, the most important manifestation of the disposal objective, was aimed at creating communities consistent with the Jeffersonian ideal of small landholdings, with the hope that this would usher in stable local economies, responsible government, and a broad range of opportunities for the individual citizen. The premises on which these expectations rested proved to be increasingly illusory as settlement moved west. Climate, soil conditions, and water supplies did not afford sufficient productivity to support family farms of the prescribed size. Additional resources were required. The Timber Culture Act of 1873 permitted applicants to obtain title to an additional 160 acres, and the similar Desert Land Act was passed in 1877. Whereas the bulk of public domain alienation occurred before the last decade of the nineteenth century, private acquisition of public lands has continued up to the present; however, since the 1930's the acreage acquired has been small.

The third era of federal resource policies and programs is that of federal management of natural resources. By the turn of the century settlement had become widespread, and the rapid shrinking of the public domain was causing alarm. Large acreages of forests were fast being depleted, and shortage of timber for future use seemed imminent. It was also feared that areas of unusual natural beauty would soon be divided and perhaps destroyed under federal land-disposal programs. It became increasingly apparent that the national interest required retention of certain portions of the public domain. The need for preservation was met first through specific legislation which set aside certain areas, and later through the establishment of agencies to administer the reserved tracts. In 1872 the first national park was established—the Yellowstone National Park. Under the Harrison Administration the first national forest was set aside. The national forest system was greatly expanded by an additional 70,000,000 acres during the first decade of the 1900's through the programs of

Theodore Roosevelt and Gifford Pinchot. Subsequent reservations of public land were made through the 1930's when Congress passed legislation requiring congressional action for future additions.

Custodial management of the reserved lands was accomplished through federal programs designed to enhance the use of publicly owned land resources. In 1903, provisions were made for wildlife refuges. Legislation in 1911 permitted the purchase of land for the specific purpose of game and wildlife management and, during the 1930's some 11,000,000 acres of submarginal agricultural land was acquired by the government for wildlife refuges. While this program was geared primarily to facilitate adjustments within agriculture, commitment of this land to wildlife was an expression of the custodial concept. The Taylor Grazing Act of 1934, perhaps the most significant single program implementing the policy of custodial federal management, established an organization to regulate the use of publicly owned grazing land by livestock ranchers.

TRANSITION IN POLICY

Policies have differed over time and between agencies in response to different interpretations of the objective that characterized each era. In the policy shifts that occurred during the third era, that of resource management, three stages may be discerned, the last of which characterizes the present policy of federal resource agencies. The growing national concern for maintaining and enhancing resource quality will, I believe, require a new orientation which will constitute a fourth stage in policy transition.

These policy orientations may be designated by identifying the domain of agency concern, the scope of their objectives, the organizations through which the objectives are pursued, and the constituencies affected by the resulting policies. One can thereby gain insight into the vehicles through which federal resource agencies strive to fulfill their management function, and one can also critically evaluate the structural shortcomings in their organization.

The first phase of federal policy toward resource management focused agency concern on individual resources for single, discrete uses representing the demands of single-interest groups. This outlook prevailed between 1890 and 1920. Rangeland was managed for grazing purposes by the General Land Office; forest lands for timber production by the newly established Forest Service; the Reclamation Service managed water for irrigation purposes in the West; the Corps of Army Engineers was concerned with flood protection and navigation. The vestiges of this policy phase are evident in some of the agencies mentioned which have retained their original constituencies. Some of these "clients" have resisted policy change through their elected representatives and through lobbying activities.

During the early decades of the 1900's, resource policy increasingly emphasized a new orientation toward the management of resources to serve several uses simultaneously. This transition involved the amalgamation of clientele groups, but it preserved the earlier pattern of agency jurisdiction. Individual agencies were considered to have primary jurisdiction over individual resources.

The concept of multiple use, popularized by the writings of Pinchot and Roosevelt, was adopted by the then infant Forest Service, which was experiencing sharp competition among the groups using national forest land. The term "multiple use" was synonymous with grazing, lumbering, and public recreation as simultaneous objectives of management. The concept was later adopted by the federal water resource agencies in their multiple-purpose projects; however, the motivation for this was due not so much to competition among different clientele groups as it was to the statutory requirements for public water project repayment, which designated each category of use as "reimbursable" or "nonreimbursable." It is hardly surprising that the federal construction agencies appropriated this notion. Their projects had to satisfy repayment requirements. Since joint costs frequently had to be allocated among reimbursable and nonreimbursable uses, a multipurpose orientation was inevitable.

The technical problems of applying the multiple-use concept are apparent: To what extent should management encompass these uses? Should use priorities be established? Some uses may be mutually exclusive; other combinations of uses may be complementary up to a given intensity threshold and competitive beyond. The emphasis on management programs that embraced several uses posed a threat to the federal agencies which had been established to provide separate functions or to serve a particular category of demand. It meant that the traditional clientele of single-purpose agencies had new avenues for expressing their demands vis-à-vis federal resource programs. The more recent suggestion that natural resources be managed in "clusters" has posed a parallel threat.

The third phase of federal policy stresses management by several agencies of particular categories of natural resources for several types of use. This emphasis was originally encouraged by a concern with the consistency and comparability of programs of different agencies that managed the same resource, as in water development. The Federal Interagency Committee Report on Benefits and Costs (1952), the Budget Bureau Circular A-47 (1953), and more recently Senate Document No. 97 (1962) attempted to establish comparable standards and criteria for the programs of the federal agencies which managed water resources. The orientation toward federal policy also tended to unify a number of previously separate user groups. For example, a vigorous national reclamation association was formed when diverse types of users, including recreation, power, and irrigation interests, were combined under the title of "western water users." This coalescence of water interests served to broaden their scope of concern with federal programs, and they have taken strong positions on the policies and programs of agencies with responsibility for resources other than water.

There are indications that a fourth stage is emerging in which policies are designed to manage several resources for a variety of uses in response to several interest groups. The coordinated efforts of several agencies are represented, for example, in the programs of the river basin commission. The objective of such plans for basic regional resources is the promotion of economic growth. It is argued that the growing concern with the quality of natural resources necessitates increased cooperation in their management.

POLICY IMPLICATIONS OF RESOURCE QUALITY

To meet the demand for higher quality in natural resources, their productivity must be maintained or increased. But how are we to interpret the productivity of a resource? Is it the *highest* current gain from a particular use? Or is it the *average* benefit derived from various uses? Or does it depend on the *number* of uses for which the resource has a positive value? The point is not to decide which interpretation is most valid but to note that the variety of interpretations poses a typical problem in the use of a productive resource. So long as there is uncertainty about the value of a resource for a particular use, flexibility is desirable so that emphasis can be shifted to other uses. Indeed, a particular "quality" objective will include elements of the relative productivity of a resource as well as its flexibility in imputing a value to that source. Although several interpretations are possible, evaluation of the qualitative aspects of resource development and management will reflect simultaneously several of these interpretations.

The important implication for policy is that a concern with the quality of natural resource must take into account a large number of economic consequences that are external to any particular production process involving that natural resource. The resource manager must be sensitive to the multiple uses of a resource and to other resources that affect its productivity in any single use. Thus the economic interaction among resources becomes important in the management process. Federal policies designed to guide the management of natural resources for "quality" objectives must be multiresource in scope.

Problems of the quality of natural resources will require the orientation we have previously identified as stage four. We find that federal policies for natural resource management predicated on growth objectives logically require the same orientation as those based upon quality objectives. This appears somewhat ironic in light of the frequent contention that qualitative objectives are different from, and usually contrary to, growth objectives. In fact, they turn out to be "sisters under the skin."

There are formidable empirical obstacles to the development of management programs embodying the concept of many resources for many uses. These obstacles are both conceptual and organizational. Conceptually, the analytical task of specifying management alternatives and evaluating them is substantial. The fact that natural resources are in part substitutable and that their interrelationships may entail technical complementarity and competition means that a large number of possible permutations will have to be investigated. Not only does this impose a heavy burden in terms of sheer numbers; it also requires research on resource interrelationships.

Organizational obstacles in the way of achieving quality objectives in resource management include the reluctance of existing agencies to surrender the jurisdictions they now enjoy.

RESEARCH: A BASIC NEED AND A PROBLEM

The increased complexity of resource problems suggests that the role of research should be given higher priority.

In the early 1960's the potential impact of river-basin development projects on local economic activities was well recognized. So, too, were the complementary relationships between the development of water and of associated resources. Recognition of these two facts led to a concern for establishing guidelines and criteria for all natural resource development within particular river basins. Applied research on each of the major river basins of the United States was initiated through a series of consortia between the four principal federal agencies for water resources development and the states involved. The preliminary findings of joint agency endeavors such as the Potomac River Basin study and the Delaware River Basin study have varied greatly in quality and comprehensiveness. The partnership in research has been handicapped by having no single agency or individual within the federal establishment responsible for these studies.

Over the last half decade, research on natural resources has been emphasized as distinct from the construction programs which dominated the older image of the line agencies. This has resulted from the growing awareness of resource interrelationships and the more explicit linkage between natural resource programs and economic growth. The potential economic leverage of annual federal expenditures in excess of three billion dollars is substantial; recent concern with area redevelopment, regional poverty, and other problems has led to programs for the development of natural resources as a means of achieving these broad social objectives. This was evidenced in the program for accelerated public works, which received a substantial congressional appropriation in 1962.

Research has been directed at two principal types of problems. The first is the implication of resource development for local economic growth. The over-all economic development plans of the Area Redevelopment Administration and the base studies of river-basin development stressed these implications. The second type of problem involves the allocation of federal funds for research and development, and requires longer-range study to provide guidance and specific criteria. Concern with this problem is suggested by a study within the Office of Science and Technology in the field of energy, and by its publication entitled *Research and Development on Natural Resources.*

The line agencies have been quick to acknowledge the implications of research activities on their construction programs. With increasingly critical positions being assumed by universities and other public groups, agencies have been pushed to prepare more adequate defenses for their proposals. These rationales require a research effort which few of the line agencies have had the competence to provide. Realizing this, the operating agencies endeavored to participate directly in such research.

AD HOC ATTEMPTS TO EFFECT
COORDINATED RESEARCH

Legislative initiatives for executive reorganization and attempts to secure coordinated resources research on an *ad hoc* basis had been made sporadically since the 1930's, but, beginning in the mid-1950's, they have followed a consistent set of recommendations patterned after those suggested by the Hoover Commission Report on government reorganization.

A bill was introduced in the 86th Congress to create a Council of Resources and Conservation Advisers in the Executive Office of the President, but failed to pass. In the absence of any mechanism for the coordination of resources research, various interagency groups have been formed to give explicit attention to particular resource problems.

During the 1960 presidential campaign, the Kennedy Administration favored the establishment of a Resources and Conservation Advisory Council in line with the earlier proposal. In February, 1961, the President's Natural Resources Message linked the analysis of natural resources problems to the planning function of the resource agencies and their construction programs. It identified the problems of fragmentation in jurisdiction and the incomparability and inconsistency in standards and criteria used in planning studies, and proposed a presidential Advisory Committee on Natural Resources as a coordinating mechanism. In April, 1961, the Administration announced specific measures to provide this mechanism. The Water Resources Planning Act which the President sent to Congress in July, 1961, would have established a council composed of the secretaries of Agriculture, Interior, Army, and Health, Education, and Welfare, and provided for river-basin commissions to conduct comprehensive land and water planning. The Conservation Message of 1962 continued this concern, but shifted major emphasis to federal research in natural resources. In response to the 1961 and 1962 commitments, the National Academy of Sciences and the Federal Council on Science and Technology undertook a study on the scope and adequacy of federal research on natural resources problems.

In addition to these studies of federal research activities, two mechanisms were created by the Kennedy Administration to deal with policy problems for particular resources. The first is the interim *ad hoc* Water Resources Council, operating at executive request on specific assignments relating to the coordination of water resources planning in major river basins by the departments of Interior, Army, Agriculture, and Health, Education, and Welfare. Staff for the Council is provided by participating departments. The second is the Recreational Advisory Council, established by executive order to serve as a focal point for the coordination of federal recreation programs. The Bureau of Outdoor Recreation, established in 1962, assisted by the participating departments, provides the Council with staff.

These and similar interagency mechanisms have facilitated the adoption of common standards and procedures to be followed by the resource agencies. Thus far, they have not undertaken to analyze the relationships between resource programs and the national economy, and therefrom to establish broad objectives for the natural resources sectors. One such attempt was made in the area of water resources; but it was under the auspices of the Senate Select Committee on National Water Resources, not under initiative of the executive branch. While the executive branch has sought, with limited success, to create badly needed machinery for handling resource management problems, existing organizations have attempted to use the avenues presently open to them. The following two examples illustrate the inadequacies of existing mechanisms for coping with present resource problems.

The energy research and development study.—The background to this study was the research and development undertaken by the Atomic Energy Commission (AEC), since the mid-1950's, on the use of nuclear power for civilian purposes. In the early years of this effort, there was an extremely high complementarity between this program and other AEC programs for military nuclear applications.

As the civilian nuclear power program became increasingly self-contained, the AEC was requested to prepare a study exploring the prospects of nuclear materials as a source of electric energy for civilian consumption. The agency made a cursory inventory of the energy resources of the United States, a relatively unsophisticated economic argument for accelerating nuclear energy development because of the "stock" nature of alternative domestic energy resources, and a proposal that it construct a series of experimental reactors. Thus the request for broad research on the future supply of and demand for energy resources ended up as a rationalization for an expanded AEC civilian program.

The Budget Bureau recognized the inadequacy of this report as a guideline for future federal policies, and asked the agency to prepare a new draft. The AEC's second study was essentially an enlargement of the first and was sent to the President. Dissatisfaction remained within the Executive Office of the President. Little consideration was given to other energy sources and to national energy surveys that were then being undertaken by the Federal Power Commission and the Bureau of Reclamation.

A study of broad scope was needed to determine criteria for budget allocations among the various federal energy agencies. The AEC effort was not of the depth required, and the Federal Power Commission and Bureau of Reclamation studies were too narrow in scope. A research team of high technical competence was needed, but its conclusions must be free from agency bias, since the report would presumably have direct bearing on budget appropriation. Who was to conduct the study? Where would it be done and under what jurisdiction? It became immediately apparent that there was no logical "home" for such a study. The executive branch decided that the President's Science Adviser and the chairman of the Council of Economic Advisers should direct the effort. An interagency steering committee was appointed, and competent technical personnel from the federal agencies as well as outside consultants were secured. The basic objective was to assess future energy demands, based on probable technological changes. The study was to evaluate research and development opportunities somewhat in the manner of benefit-cost analysis to ascertain a priority ranking of problems deserving federal support.

No final report from this study has yet been made public and its competence and completeness cannot be assessed at the present time. The scope of the investigation was correctly perceived, however, in light of the total federal effort in the field of energy resources. The difficulty of not having a readily available staff was overcome to some degree by borrowing from the action agencies. Vesting the chairmanship of the study committee in the Office of the Science Adviser and the Council of Economic Advisers minimized the possibility of introducing agency bias.

The Petroleum Study Committee.—A study was undertaken in 1961, under the auspices of the Office of Emergency Planning, to determine the effects of the oil import quota system. The ensuing report strongly suggested that the quota system and rules governing the exchange of quota tickets—permits to import a designated amount of foreign-produced petroleum—increased the cost of petroleum and petroleum products to domestic consumers. The Committee recommended several specific studies to consider changes in existing policies, but none of these were activated.

A task force group was established within the Office of Statistical Standards of the Budget Bureau to investigate the adequacy of the statistics on petroleum reserves and cost of production. The group has made recommendations for augmenting and improving domestic data.

These two *ad hoc* efforts to cope with research on national resource policies illustrate the serious defects in the organizational machinery available for basic research on resource management. (1) No clear structure of authority was provided to assure high technical competence and freedom from bias. (2) No mechanism was provided for the continuing research that is needed. (3) The structure of the research group did not permit a clear statement of the policy implications of the technical studies.

In the energy study, a one-shot effort was made to assess the nature and extent of current research and development and to pinpoint the areas that promise to reduce cost or increase energy output. The problem was conceived as a static one, however, and the ensuing report does not provide for a continuing survey of ways in which consumption and demand patterns for energy may change in the future.

The petroleum example illustrates how an initial study, designed to produce research results relating to an area of real concern to policy, was broken up. The resulting study by the Office of Statistical Standards does not propose a research entity that could relate the new and improved statistics to the policy issue.

THE PROSPECTS FOR ORGANIZATIONAL IMPROVEMENT

If natural resources are to be managed in conjunction with plans for economic growth or broad qualitative objectives, the programs and policies of the various action agencies must be coordinated. To achieve this, planning must be comprehensive enough to incorporate the programs of all the agencies concerned. Planning on a broad scope can best be accomplished in a single, central unit, which would provide guidelines for the policies and programs of the operational agencies.

The functions of this resource analysis unit may be considered in two major categories. The first includes the following functions: (a) the identification of relevant problems for analysis; (b) the acquisition of adequate and timely data; (c) the competent performance of the research and analysis these problems involve; (d) the development of analytical methods and procedures that are relevant for the analysis of resource policies.

A second category of functions is needed if the results of the analytical unit are to provide a basis for resource policy: (a) the assessment of the implications of such analysis for existing programs and policies; (b) the making of this information available to resource agencies and to the public; (c) the utilization of the information within the decision-making process.

Performance of these functions requires certain properties or conditions within the analysis unit:

1. A broad perspective must be established and maintained. The scope of concern must include all natural resources so that their interrelationships may be considered in the formulation and analysis of relevant problems. Such a scope has been approached at the regional level in the development of plans for river-basin development, but it is not broad enough, in terms of the resources or the geographical areas considered, for the functions identified above.

2. Long-run shifts in resource supply and demand and their relation to economic growth must be considered if federal research, development, and management are to elicit the greatest contribution from our natural resources. More research is needed on the timing of resource planning and management.

3. Multidisciplinary skills are needed in the formulation of policies, and the interrelation among the physical, biological, and social sciences must be more clearly understood.

4. There should be access to both governmental and nongovernmental analytical skills, facilities, and data.

5. Specification of the research problem, selection of data, and interpretation of the analytical results should be objective and free from bias.

One of the obstacles in achieving a research analysis unit is the unwillingness of the resource agencies to create a superauthority for planning, whether it be a department or an office under an independent authority. Another obstacle is the unwillingness of Congress to relinquish its traditional political role in specifying the alternatives for resource programs.

The first obstacle may be likened to the difficulties encountered in proceeding from an oligopoly to a monopoly. The executive agencies have proceeded in a *quid pro quo* pattern in the past. Their relationships and alliances within the executive branch and with Congress have been predicated on this *modus operandi*. A new pattern of deciding what needs to be done and who will do it holds the threat of uncertainty for individual agencies. New lines of communication, bargaining, and mutual reinforcements would be required to protect and perpetuate agency interests.

Similarly, Congressional objection stems from the threat of losing a historical position as initiator of policies for federal resource development. With the important exception of agriculture, legislative committees, including the substantive and appropriation bodies in both houses, have initiated federal policies on natural resources, thus reversing the traditional "proposing" and "disposing" functions of the executive and legislative branches.

One significant distinction between the executive resource agencies dealing with resources and the Department of Agriculture has been the strong research tradition of the latter. Even before the organization of the Bureau of Agricultural Economics, the Department fulfilled research functions greater in scope and with a more adequate technical staff than was true of the Department of Interior or other resource agencies. This strong research arm led to an intradepartmental analysis of problems and possible solutions that culminated in strong proposals for national policy. Other resource agencies, lacking the tradition, the proficiency, and the reputation for research of high professional quality, were handicapped in this regard.

Furthermore, there was greater legislative interest in resource development programs than in agriculture. These programs meant brick-and-mortar projects with their immediate impact on local employment and prospects for tangible monuments to the beneficence of local representatives. This strong motivation for control over programs and policy initiatives by the legislative branch and the increasing competitiveness among the executive resource agencies led to an accumulation of power in the legislative domain. Paralleling this shift, the office of Secretary of the Interior has become less effective in executive branch coordination for national resource policies. Indeed, since the 1930's this function has been increasingly taken over by the Budget Bureau. This unit, however, in the capacity of "broker" for all administration policies, is not structured or staffed to perform this task for the natural resources sector.

The inevitable result has been alliances between the individual resource bureaus and agencies and the legislative committees. The resulting proposals have often been initiated by the legislative bodies, fitted into the mission-oriented rationale of the resource agencies, and forwarded to the Budget Bureau. At this juncture the Budget Bureau attempts to transform the Administration's proposals into legislation. In the process there is substantial *quid pro quo* "trading," during which many of the original proposals may be scrapped. The important point, however, is that no coordinated set of proposals is considered and, of even greater importance, no overall guide for integrated development of resources emerges.

While there are no indications that a central analysis unit will be established in the immediate future, several concrete steps recently taken show an awareness of the need for comprehensive, coordinated planning to deal with the problems of natural resources quality.

These changes effectively broaden the scope of research in two important areas.

Evidence of the first type of change may be found in the Bureau of Outdoor Recreation, initially established in 1961 to provide a "secretariat" for the President's Recreational Resources Council. This Council, parallel in structure to the Water Resources Council, was comprised of the secretaries of the four resource departments and reported directly to the President. Staff for the agency was to be provided by the Bureau of Outdoor Recreation, housed in the Department of the Interior, but staffed by all departments represented on the President's Recreational Resources Council. Ambiguity in the wording of the executive order establishing the office made it uncertain whether the Bureau was to become an integral organ of the Interior Department, a multidepartmental entity reporting to the Council, with secondary responsibilities to all participating departments.

Some of the original ambiguity has been clarified. The Bureau's budget became a separate item in the 1965 budget, and its staffing has proceeded independently of the Department of Agriculture, the Corps of Army Engineers, and the Department of Health, Education, and Welfare. The Bureau of Outdoor Recreation seems to be evolving into an integral part of the Department of the Interior. Having no physical programs, the Bureau has been oriented toward the Office of the Secretary. Thus in its research function it will deal with problems relevant to the entire Department of the Interior. It remains to be seen whether it will make recommendations on the management of individual resources, such as water or grazing, which would unify the impact of various resources on recreation.

A parallel development is the recent passage in the House of the Watershed Planning Act, which legitimizes the *ad hoc* Water Resources Council and provides a basis for the Council to assemble its own staff. Here again is the possibility of establishing research competence adequate to promulgate policies and guidelines for problems of resource quality.

If this trend is continued, the executive branch should be able to initiate resource policies and programs that take explicit account of quality objectives. Federal policies for natural resources would then be more closely in line with those for agriculture, restoring the traditional function to the executive branch of the federal government.

Simply to criticize existing policies and programs for their failure to stress the quality of natural resources is neither meaningful nor constructive. The present resource agencies are not well constituted to perform the research and planning needed to achieve a comprehensive program. The Administration appears to be building up research and planning competence in several interagency organizations. While this approach will help the federal establishment deal with qualitative problems, it does not seem fully adequate for coping with problems involving key programs in competitive departments. Problems of this type require an authority with super-cabinet status. Although other demands prevent the President from giving these issues more than a small percentage of his time, their political leverage is high—perhaps sufficiently so that these decisions will always remain a Presidential function.

INDEX

Adiabatic systems, 24–25
Agricultural technology: effects on ecology, 65–70. *See also* Soil
Agriculture, Department of: resource research, 210
Air pollution: public health aspects of, 113–120; acute effects, 113–114; disasters, 113–114; smog, 114, 115; chronic effects, 115; levels of, 115–117; gaseous contaminants, 117; California standards of ambient air quality, 115, 116, 117; air-pollution control programs, 118–120; Clean Air Act of 1963, 118; alert stages for toxic pollutants, 119; atmospheric pollutants, 121; biochemical aspects of, 121–126; ozone, 121, 122, 123, 126; peroxyacetyl nitrates, 121–126 *passim;* lethal doses, 122; effects on plant life, 122–123; effect on carbohydrate metabolism, 123; action on enzymes, 123–126
Atomic Energy Commission civilian research program, 206–207
Atmospheric pollutants. *See* Air pollution

Barlow Commission, 194–195
Biochemical aspects of air pollution, 121–126
Biology: taxonomy in, 22
Biosystem, 25–27. *See also* Ecosystem
Bureau of Outdoor Recreation, 211
Burning, prescribed: in wildlife management in California, 71–83; for ponderosa pine, 76–78; for chamise chaparral, 80–81; for woodland-grass vegetation, 82–83

Carbohydrate metabolism: effect of air pollutants on, 123
Carbon monoxide as air contaminant, 117
Carrying-capacity concept in resource management, 155
Chamise chaparral: fire ecology in, 79–81
City planning: factors in utopian approach, 182–186; problems in the United States, 193–194. *See also* Planning
Clean Air Act of 1963, 118
Community, ecosystem, and population, definitions, 140, 141, 147–148
Conservation: new urban emphasis on, 165–166; benefits and costs of, 166–169. *See also* Resource administration; Resources, natural
Cybernetics concept in ecosystem research, 150

Decision-theory concept in ecosystem research, 151
Deductive planning, 177–182, 191–196
Demineralizing processes in water purification, 104–111. *See also* Water
Desalination methods, 104–111. *See also* Water
Distillation method of water purification, 104, 108

Ecology: disruption by mass agricultural production techniques, 65–70; definition, 145n; of ecosystems, 145n, 145–146. *See also* Ecosystem

213

Lightning Source UK Ltd.
Milton Keynes UK
UKHW010108070123
414950UK00004B/296